Medical Laboratory Technology

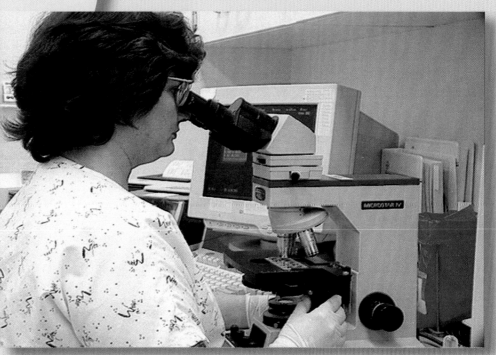

SCIENCE in a technical world

A Project of the Education Division of the American Chemical Society

 *W. H. Freeman and Company
New York*

Science in a Technical World Project Team
Co-Principal Investigators: Sylvia Ware, David Lavallee, Kenneth Chapman
Project Director: Ann Benbow
Production Director: Colin Mably
American Chemical Society: Sylvia Ware, Ann Benbow, Janet Boese, Gary Dikeos,
    Helen Herlocker, Guy Belleman, Michael Tinnesand
Writers: Jack Ballinger, Ann Benbow, Helen Herlocker, Wayne Morgan, Carol Muscara,
    Marti Tomko, Emmett Wright, Robert Bernoff, Ken Weese
Editors: Marcia Vogel, Ann Benbow, Colin Mably
Project Evaluator: Robert Bernoff
Assessment Consultant: Gail Goldberg
Video Coordinator: Colin Mably
Video Production: Take One Video
CD-ROM Development: CTS, Inc., Sonalysts, Inc.
Safety Reviewer: W. H. Breazeale
Content Advisor: Andrea Bennett

National Science Foundation Visiting Committee
Carlo Parravano, Chair; Barbara Wrobleski, Michael Farmer, Mickey Sarquis,
Nancy Dillon, Robert Maleski, Felicia Graves, Toby Horn

W. H. Freeman and Company Book Team
Publisher: Michelle Russel Julet
Sales and Marketing Consultant: Arthur C. Germano
Project Editor: Mary Louise Byrd
Text Designer: Blake Logan
Cover Designer: Patricia McDermond/Blake Logan
Cover Illustration: Roy Weiman
Illustration Coordinator: Lou Capaldo
Illustrations: Proof Positive/Farrowlyne Associates, Inc.
Photo Research: Proof Positive/Farrowlyne Associates, Inc.
Production Coordinator: Julia DeRosa
Production and Composition: Proof Positive/Farrowlyne Associates, Inc.
Manufacturing: RR Donnelley & Sons Company

Library of Congress Cataloging-in-Publication Data

Medical laboratory technology : a project of the Education Division of the American
    Chemical Society.
        p. cm. — (Science in a technical world)
   ISBN: 0-7167-4033-8
   1. Medical laboratory technology—Vocational guidance. I. American Chemical
Society. Education Division. II. Series.

RB37.5 .M425 2001
616.07'56'023—dc21   2001033567

Development of the *Science in a Technical World* program was supported, in part, by
the National Science Foundation under Grant Nos. DUE 9454564 and DUE 9752102.
Opinions expressed are those of the authors and not necessarily those of the
Foundation.

Printed in the United States of America
First printing 2001

Science in a Technical World
Industry Contacts

ROBERT HOFSTADER (industry consultant); TERRY WEBB, Quality Assurance, Pepsi Bottling Group; JEFF FLYNN, Kleentek; RONALD BARBARO, Northern Virginia Community College; STEVEN CAWTHRON, Leesburg Wastewater Treatment Facility; KENNETH GLANZ, Appleton Papers, Inc.; MARIA SPINU, DuPont Research and Development; PATRICIA LISTER, DuPont Marshall Laboratories; PETER BICK, Eli Lilly and Company; MARK A. HOLCOMB, Dominion Semiconductors; PAMELA MCCARTHY, Civista Medical Laboratory; JENNIFER FERNS, Federal Bureau of Investigation; JERGEN VON BREDOW, Food and Drug Administration; FRED CONFORTH, Phillips 66 Company

Science in a Technical World
Pilot Teachers

JEFF ALISAUCRAS, South Carroll High School, Sykesville, MD

JAMES BAUER, McKinley High School, Honolulu, HI

BRITTANY BEHRENS, Coronado High School, Scottsdale, AZ

BARBARA BILLINGS, Milton High School, Alpharetta, GA

BILL CALLAHAN, Watsonville High School, Watsonville, CA

JIM DIXON, Silver Lake Regional High School, Kingston, MA

KATHLEEN DOMBRINK, McCluer North High School, Florissant, MO

RONALD DUNAWAY, Monteray High School, Lubbock, TX

JONATHAN DUNSKI, North Carroll High School, Hampstead, MD

DOUG ENGLAND, Wallace High School, Wallace, ID

ROBERT FOUR-HOGUE, South Carroll High School, Sykesville, MD

JACK GOLDSTON, Ross High School, Fremont, OH

ALAN GONZALES, Lakewood High School, Lakewood, CO

KRISTA GRAY, McDowell Senior High School, Erie, PA

BERNIE HERMANSON, Harlan High School, Harlan, IA

NICK HOFFMAN, Wallace High School, Wallace, ID

JOCELYN HOLLIS, Chester High School, Chester, PA

SAMUEL HOLTZMAN, Pilgrim High School, Warwick, RI

JULIA JANOWICH, Francis Scott Key High School, Union Bridge, MD

CINDY JONES, Jefferson Senior High School, Edgewater, CO

LEAH LINDBLOM, Arvada West High School, Arvada, CO

MARY LINDIMORE, Arvada West High School, Arvada, CO

KAREN LUNIEWSKI, North Carroll High School, Hampstead, MD

DIANNE MCCLEARY, Montwood High School, El Paso, TX

TOM MCCOY, Keystone High School, Knox, PA

SUE MATTESON, McDowell Senior High School, Erie, PA

SHARON MUENCHOW, Arvada West High School, Arvada, CO

ELLIE MUNIZ, Ridgeview High School, Columbia, SC

MIKE MURRAY, Silver Lake Regional High School, Kingston, MA

MARK NALE, Tyrone Area High School, Tyrone, PA

KRISTEN NELSON-MCKENZIE, Thornridge High School, Dolton, IL

BEN NORTON, Bergen County Technical High School, Hackensack, NJ

SCOTT OLSON, Red Rocks Community College, Lakewood, CO

MARIA PACHECO, Buffalo State College, Buffalo, NY

ROBIN PARKINSON, Northridge High School, Ogden, UT

SHELLY PERETZ, Thornridge High School, Dolton, IL

RUSS RAPOSE, Pilgrim High School, Warwick, RI

CANDY RUZIECKI, Chester Senior High School, Chester, SC

BARBARA RYDEN, McDowell Senior High School, Erie, PA

THOMAS SANDHAM, North Kingston High School, North Kingston, RI

WAYNE SILL, Hutchinson High School, Hutchinson, KS

GEOFF SMITH, Watsonville High School, Watsonville, CA

BILL SPEED, Need School, Kailua, HI

TONY TAYLOR, Jefferson County Open School, Lakewood, CO

KATHLEEN THOMPSON, Anthony Jr/Sr High School, Anthony, TX

SCOTT THOMPSON, Arvada High School, Arvada, CO

LILLIE TUCKER-AKIN, Tupelo Christian School, Tupelo, MS

BARBARA WALKER, Ottumwa Community Schools, Ottumwa, IA

LARRY WEATHERWAX, Mclaughlin School, Alaska

KEN WEESE, Montwood High School, El Paso, TX

GARY WHITLING, Keystone High School, Knox, PA

HAROLD WILSON, Medomak Valley High School, Waldoboro, ME

DELORES WRIGHT, Chester High School, Chester, PA

DOUG YUST, Francis Scott Key High School, Union Bridge, MD

Teacher Review Panel

Bernie Hermanson

Nick Hoffman

Mark Nale

Candy Ruziecki

Barbara Walker

Larry Weatherwax

Ken Weese

contents

Introduction   1

The Problem   1
Your Role in Solving the Problem   1
Overview of the Workplace: Fact Sheet   3
   on Medical Laboratory Technology

Technician Orientation   5

Developing Your Skills Base   5
Important Concepts, Processes, and Skills   6
Step 1: Setting Up and Maintaining a Laboratory Notebook   7
Step 2: The Impact of Automation in the Clinical Laboratory   9
Step 3: Introduction to the Clinical Laboratory   13
Science Connections: The Blood   15

The Work   21

Laboratory 1: Blood Banking   21
Science Connections:
   Blood Bank Products   22
Science Connections: Human
   Blood Types   23
Laboratory 2: Blood Cells   28
Laboratory 3: Clinical Chemistry   39
Science Connections:
   Standards, Controls, and
   Samples   41

☼ Science Connections: Blood Sugar and Diabetes 42

☼ Science Connections: The Kidneys 44

Laboratory 4: Bacteriology 45

☼ Science Connections: Gram Stains and Medicine 46

Laboratory 5: The Night Shift 51

Data and Results 53

Presenting your Findings 53

Glossary 55

Additional Resources 59

Credits 60

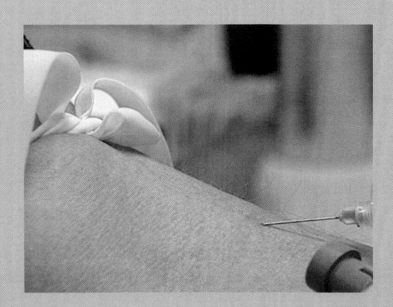

Multimedia Resources

Medical Laboratory Technology Video: This 16-minute video takes you inside a hospital laboratory where technicians run many different tests and talk about their work.

Medical Laboratory Technology CD-ROM: This CD-ROM is organized into sections. Medical Laboratory Technology Overview introduces you to the general role of clincal laboratory technicians. Hospital Laboratories give you an overview of the different tests that are run by technicians. Technician Tasks gives specific information about each of the laboratory tests and processes that technicians are responsible for in a hospital laboratory. Practice gives you an opportunity to practice what you have learned about medical laboratory procedures. Assessment helps you assess your knowledge and skills relating to medical laboratory procedures. Key terms and definitions related to medial laboratory procedures and technician tasks are in the CD-ROM's *Glossary* and *Encyclopedia*.

Medical Laboratory Technology on the Internet: A list of Web sites related to Medical Laboratory Technology, which you will find useful throughout your work in this module, is included at the back of the book.

www.whfreeman.com/STW Additional resources for instructors and students are available online at the *Science in a Technical World* support Web site.

Science Connections Science Connections boxes offer background information and explanations to enhance your understanding of important science concepts covered in this module. You will find them near the places in the book where a key science concept is first introduced.

Science in a Technical World modules are intended for use by students in the classroom laboratory under the direct supervision of a qualified science teacher. The experiments described in this book involve substances that may be harmful if they are misused or if the procedures described are not followed. Read cautions carefully and follow all directions. Do not use or combine any substances or materials not specifically called for in carrying out experiments. Other substances are mentioned for educational purposes only and should not be used by students unless the instructions specifically so indicate.

The materials, safety information, and procedures contained in this book are believed to be reliable. This information and these procedures should serve only as a starting point for good laboratory practices, and they do not purport to specify minimal legal standards or to represent the policy of the American Chemical Society. No warranty, guarantee, or representation is made by the American Chemical Society as to the accuracy or specificity of the information contained herein, and the American Chemical Society assumes no responsibility in connection therewith. The added safety information is intended to provide basic guidelines for safe practices. It cannot be assumed that all necessary warnings and precautionary measures are contained in the document and that other additional information and measures may not be required.

welcome!

What Are the Goals of *Science in a Technical World*?

Before you begin working with this module, it is important for you to understand what the entire *Science in a Technical World* program is set up to accomplish. Each *Science in a Technical World* module is designed to help you to

- gain an understanding of the **science technology** used in a variety of workplaces and to become skilled in using science technology to solve problems that actually occur

- become knowledgeable about the **technologies** in general and compare the employment opportunities that a number of science technology workplaces offer

- work as a team in a **virtual workplace** setting

What Is Science Technology?

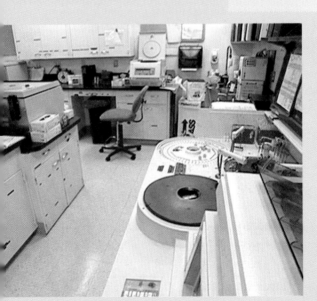

Science technology may be a term that is unfamiliar to you. It refers to the applied science that is used in a wide variety of settings, the medical laboratory being one.

For example, science technology includes the laboratory equipment and instruments used to monitor patient care. These can be the same kind of "wet lab" equipment you may have used in other high school science courses (such as chemistry, biology, or the earth sciences), or they may be very sophisticated, computerized instruments.

This program will give you the chance to learn science technology by letting you and your classmates solve actual problems from different industries and workplaces.

What Is the Technical Workplace?

The technical workplace is a laboratory setting in which technicians apply their skills to solving problems. Many of these workplaces are industries. **Industry** is the process by which products are made. Some industries with which you may be familiar include the automotive industry (manufactures cars), the food industry (processes food products), the pharmaceutical industry (makes medications), the paper industry, and a number of others.

Science in a Technical World problems may come from these familiar industries, as well as other workplaces in which technicians perform tests to provide information for solving nonindustrial problems—workplaces like a forensics or medical technology laboratory. All workplaces chosen for the entire program allow you to gain a wide variety of technical skills and knowledge in biology, chemistry, physics, and earth science.

What Is the Virtual Workplace?

The **virtual workplace** is a term used in the *Science in a Technical World* program to describe a new way of thinking about your classroom, your teacher, and your classmates. As much as possible, the work that you do throughout the program will mirror what actually happens in industry. This means you will work as teams to solve problems and will be dependent on the work of other teams, at times, to achieve results. You will do research to find information from a variety of sources (including a CD-ROM, a video, and the Internet) to help you solve problems. You will work according to a timeline that all of you establish. You will also keep very careful records of your work, which you will need to have witnessed and signed off on a regular basis. Your teacher's role will shift from an instructor to more of a manager or supervisor of your work.

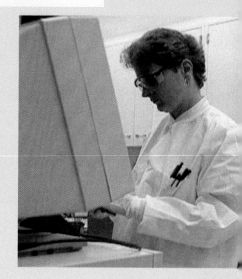

What Happens After *Science in a Technical World*?

You may decide, after completing the *Science in a Technical World* program, that you would like to learn more about the particular technologies that are the focus of the different modules. There are a number of programs that prepare people to work in science technology professions. You might want to investigate these, or contact employers directly to find out what training opportunities they offer.

Whatever you decide, *Science in a Technical World* will give you the knowledge and skills to make an informed choice about careers you may never have considered before. It will also give you a wide breadth of skills that you can use in other science-related fields.

Now, welcome to *this* module, which focuses on the workplace of the medical technician.

watch for these symbols

Look for these special symbols as you are reading this book. They have been included to alert you to important learning opportunities, activities, and safety precautions.

? Key Question Consider Key Questions carefully. These questions are designed to help you to build your understanding of the concepts, processes, and skills that are central to the focus industry of this *Science in a Technical World* module.

Tips and Hints Tips and Hints offer ideas that will help you along the way in your investigations. They may assist you in answering questions, setting up laboratory equipment, solving problems, or pulling ideas together.

note Notebook Entry This symbol reminds you to stop and make an entry in your laboratory notebook. This could be a record of a laboratory procedure, data, an answer to a question, a conclusion, or other important information.

Boldface Terms Definitions are available for terms set in boldface type in the text. They can be found in the Glossary at the end of the module.

MULTIMEDIA SYMBOLS Each *Science in a Technical World* module includes an introductory video, a multipurpose CD-ROM, and suggestions about industry information that can be found on the Internet. The following symbols alert you to opportunities to supplement your laboratory and classroom activities with one of these multimedia resources.

CD-ROM This symbol tells you to go to the module's CD-ROM and access one or more of its sections. The CD-ROM might include an Introduction, Technician Tasks, Practice, Assessment, and Reference sections.

Video When you see this symbol, you can watch the video to learn more about the production process of the module's focus industry and the role played by technicians in that process.

Internet This symbol indicates that information that will help you in your investigations is available by accessing the Internet.

SAFETY SYMBOLS   Safety precautions are extremely important in laboratory work. Be alert to the safety symbols associated with each investigation in this module, and always follow the safety rules they represent.

Safety Goggles   This symbol reminds you to wear safety goggles during the laboratory investigation to protect your eyes from chemical splashes or sudden impacts.

Laboratory Apron   This symbol reminds you to wear a laboratory apron during the investigation to protect your clothing and skin.

Hot Object or Flame   When you see this symbol, be alert to occasions when you might encounter a hot object (such as a hot plate) or an open flame during the investigation. Secure loose sleeves and long hair before starting your laboratory work.

Poisonous Substance   This symbol warns that you will be working with substances that should not be tasted. No substance should ever, under any circumstances, be tasted in a classroom laboratory setting.

Special Disposal   This symbol indicates that a substance you will be working with must be disposed of in a special way. You will receive directions on how to dispose of the substance correctly. It is important for safety, health, and environmental reasons that you follow these directions exactly.

Dangerous Vapors   This symbol warns you that dangerous vapors might be released in the course of the laboratory investigation. Whenever you see this symbol, your investigation should be conducted under an operating fume hood.

Hygiene   This symbol reminds you to wash your hands carefully after a procedure or at the end of the laboratory, as directed.

using your multimedia

Each *Science in a Technical World* module comes with its own video program and CD-ROM.

The video program takes you on a tour of the module's focus workplace. You will get to see what a medical technology laboratory looks like, the kinds of equipment available, and the way that testing materials are organized. The video also shows what technicians do in the laboratory, so you can compare their work with what you do in the laboratory. You may see the complete video at the beginning of the module, or your teacher may show certain parts at appropriate times.

The CD-ROM is a resource for you to use throughout the module. You can access its sections from the main menu by following the voice directions and clicking on the proper screen icons.

The CD-ROM has the following sections:

The **Overview** provides you with information about the focus workplace, its products, its processes, and its technicians.

Technician Tasks gives information about what technicians do in the workplace. These tasks can include sample preparation, sample checking, sample storage, testing samples, and recording results.

Practice Exercises gives you an opportunity to simulate some of the tests that technicians do in the workplace, expand your knowledge of terms and equipment, and manipulate data sets.

start

File Sound Reference Notebook Print Go History Help

Medical Laboratory Technology

Main Menu

Overview
Hospital Laboratories
Technician Tasks
Practice
Assessment

Assessment allows you to test your knowledge and proficiency of the tests, terms, procedures, equipment, and calculations that technicians need to know in the workplace.

Pull-down menus and icons launch CD-ROM tools:

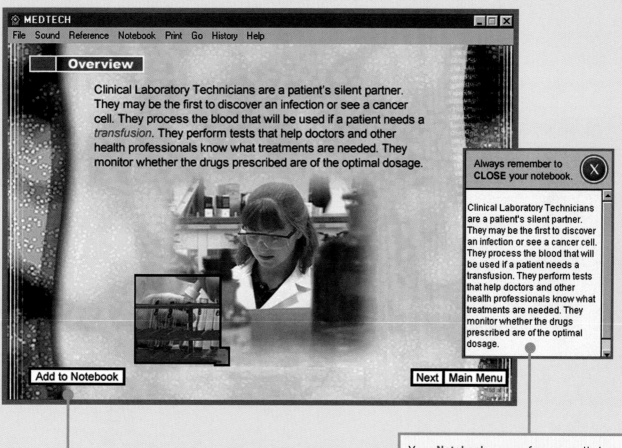

Clicking **Add to Notebook** will copy the text from the current screen to a special CD-ROM notebook.

Your **Notebook** opens from a pull-down menu. You can edit, save, and print the contents of your notebook. Your teacher may ask you to use your notebook when you are doing the Assessment portion of the CD-ROM.

Watch for **Multimedia Links** throughout this book. They suggest what part of the video or CD-ROM will be most useful to you at each stage of your work with this module.

History tracks where you've been on the CD-ROM. You can navigate to a previous screen by clicking on its name in a displayed list of screens.

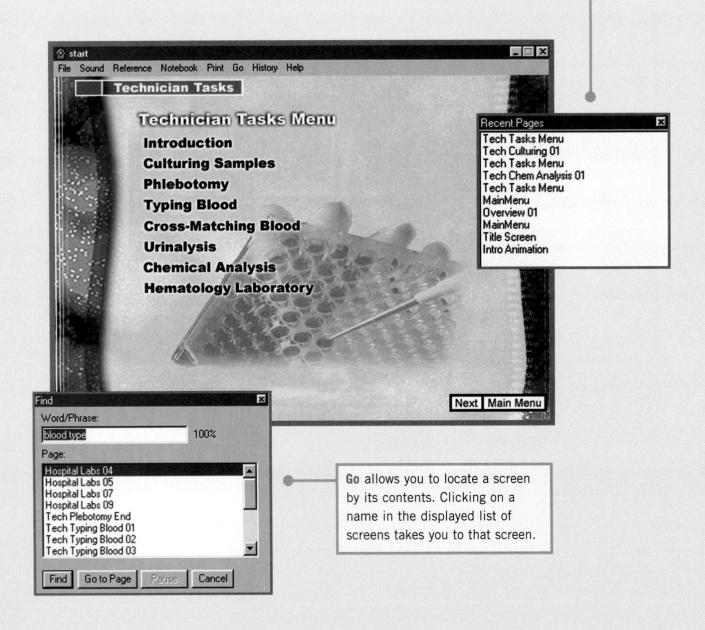

Go allows you to locate a screen by its contents. Clicking on a name in the displayed list of screens takes you to that screen.

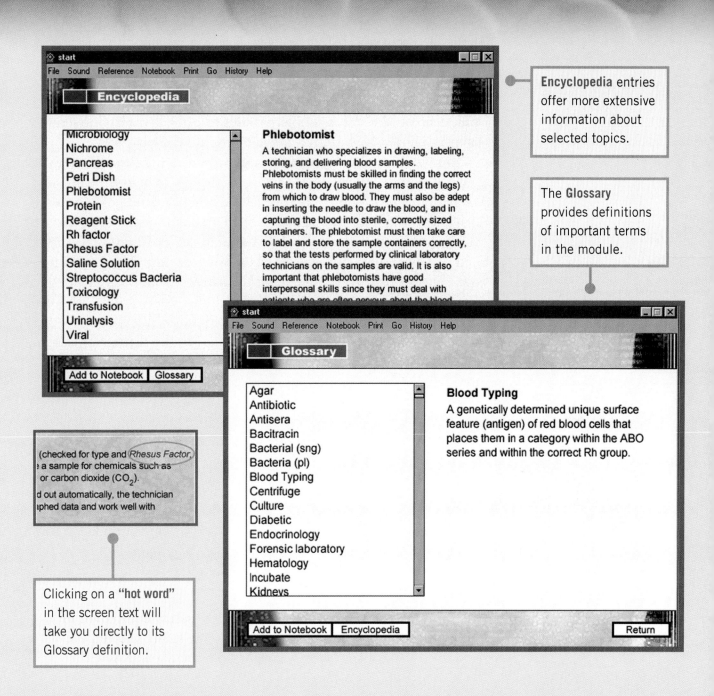

Encyclopedia entries offer more extensive information about selected topics.

The **Glossary** provides definitions of important terms in the module.

Encyclopedia

Microbiology
Nichrome
Pancreas
Petri Dish
Phlebotomist
Protein
Reagent Stick
Rh factor
Rhesus Factor
Saline Solution
Streptococcus Bacteria
Toxicology
Transfusion
Urinalysis
Viral

Phlebotomist

A technician who specializes in drawing, labeling, storing, and delivering blood samples. Phlebotomists must be skilled in finding the correct veins in the body (usually the arms and the legs) from which to draw blood. They must also be adept in inserting the needle to draw the blood, and in capturing the blood into sterile, correctly sized containers. The phlebotomist must then take care to label and store the sample containers correctly, so that the tests performed by clinical laboratory technicians on the samples are valid. It is also important that phlebotomists have good interpersonal skills since they must deal with patients who are often nervous about the blood

Add to Notebook   **Glossary**

Glossary

Agar
Antibiotic
Antisera
Bacitracin
Bacterial (sng)
Bacteria (pl)
Blood Typing
Centrifuge
Culture
Diabetic
Endocrinology
Forensic laboratory
Hematology
Incubate
Kidneys

Blood Typing

A genetically determined unique surface feature (antigen) of red blood cells that places them in a category within the ABO series and within the correct Rh group.

Add to Notebook   **Encyclopedia**   **Return**

(checked for type and *Rhesus Factor,*) a sample for chemicals such as or carbon dioxide (CO_2).

d out automatically, the technician aphed data and work well with

Clicking on a **"hot word"** in the screen text will take you directly to its Glossary definition.

Watch for **Multimedia Links** throughout this book. They suggest what part of the video or CD-ROM will be most useful to you at each stage of your work with this module.

safety in the laboratory

In the technical workplace, safety is extremely important. Laboratories rely on health and safety guidelines set by OSHA, the Occupational Safety and Health Administration of the U.S. Department of Labor. These guidelines are tailored to particular processes and practices. In some laboratories, for example, visitors to a facility must undergo a safety orientation before they are allowed into the work area.

You will have the opportunity and responsibility throughout your *Science in a Technical World* work to observe standard health and safety practices in the laboratory. These practices and guidelines are for your own good; they will ensure that your investigations are rewarding and safe experiences for you and your classmates.

Each laboratory in this book has a Safety section. It is very important that you read and understand the safety precautions before you begin any lab. If you don't understand the safety rules, ask your teacher to explain them. The Rules of Laboratory Conduct listed below apply to *all* laboratory activities included in the *Science in a Technical World* program. Read them now and review them periodically until they become second nature.

Rules of Laboratory Conduct

- Perform laboratory work only when your teacher is present. Unauthorized or unsupervised laboratory work is not allowed.

- Your concern for safety should begin *before* you start a laboratory activity. Always read and think about each assignment before starting it. Ask your teacher to explain any safety rules you don't understand.

- Know the location and correct use of all safety equipment in your laboratory. These should include the safety shower, eye wash, first-aid kit, fire extinguisher, and blankets. Know the location of all exits and proper evacuation routes.

- Wear impact/splashproof safety goggles for all laboratory work.

- Make every effort to protect bare skin. Shorts or short skirts must not be worn; long pants and shirts with sleeves provide better protection. Wear closed shoes (rather than sandals or open-toed shoes).

- Restrain loose articles of clothing, jewelry, and hair.

- Clear your laboratory bench of all unnecessary materials, such as books or articles of clothing, before starting your work.

- Be extremely careful when using sharp tools; never handle their cutting edges, and be sure to store them in an appropriate covering or case when not in use. Also, handle laboratory glassware with care to avoid breakage and the danger of getting cut.

- Check chemical labels twice to make sure you have the correct substance and the correct concentration of a solution (some chemical formulas and names can differ by just a single letter or number). Read and follow the appropriate safe-usage guidelines for each chemical.

- You may, at times, be asked to transfer laboratory chemicals from a common bottle or jar to your own test tube or beaker. Do not return excess material to its original container unless authorized by your teacher.

- Avoid unnecessary movement and talk in the laboratory.

- Never taste laboratory materials. Do not bring food, drinks, or chewing gum into the laboratory. Do not put fingers, pens, or pencils into your mouth while in the laboratory.

- Wash your hands before leaving the laboratory!

- If you are instructed to smell something, do so by fanning some of the vapor toward your nose. *Do not* place your nose near the opening of the container. Your teacher will show you the correct technique.

- Never look directly down into a test tube or beaker; view the contents from the side. Never point the open end of a container toward yourself or a neighbor. Never heat a container directly in a Bunsen burner flame.

- Any laboratory accident, however small, should be reported immediately to your teacher.

- If a chemical spills on your skin or clothing, quickly rinse the area with plenty of water. If the eyes are affected, washing with water must commence immediately and continue for a minimum of 15 minutes or until professional help arrives.

- When discarding or disposing of used materials, carefully follow the instructions provided.

- Return equipment, chemicals, aprons, and protective goggles to the designated locations.

- If in doubt about a procedure or precaution, *ask!*

- Disposable protective gloves must not be reused. The used gloves must be properly discarded.

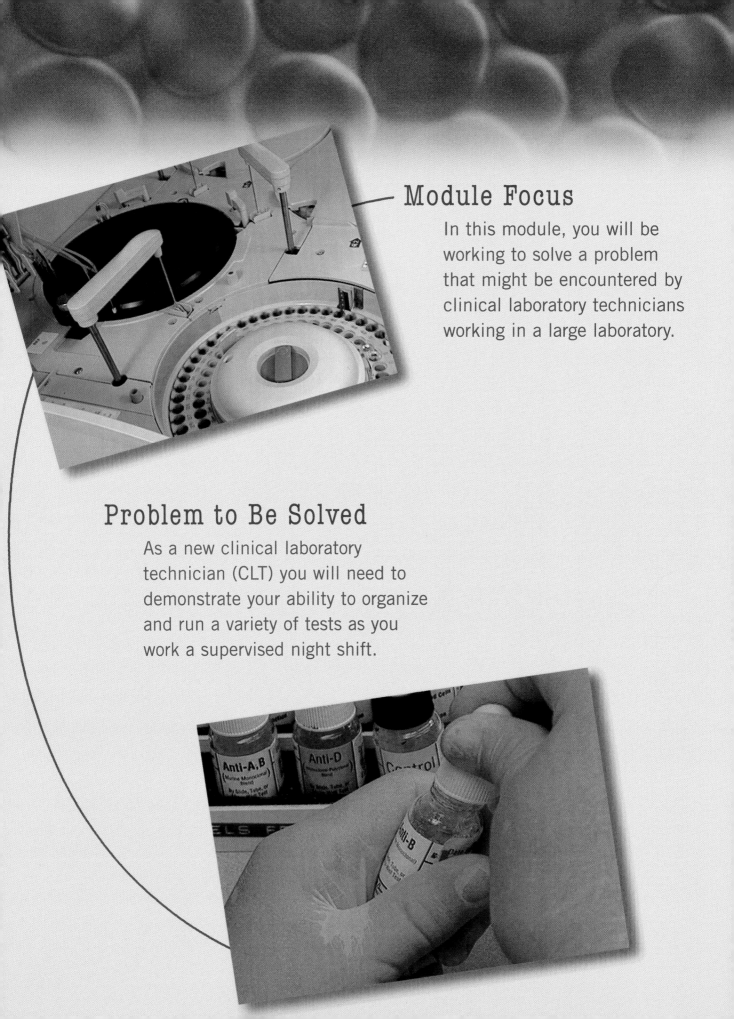

Module Focus

In this module, you will be working to solve a problem that might be encountered by clinical laboratory technicians working in a large laboratory.

Problem to Be Solved

As a new clinical laboratory technician (CLT) you will need to demonstrate your ability to organize and run a variety of tests as you work a supervised night shift.

introduction

The Problem

You are a new technician assigned to the overnight shift in a large clinical laboratory. As part of your on-the-job training as a new technician, you are learning to

- *run* laboratory tests
- *prioritize* your work
- *process* specimens for the day shift
- *perform* quality control tests, and
- *report* your results quickly and accurately.

This training is to prepare you for the range of tasks you may encounter on your shift. You will learn to process a wide variety of specimens, which sometimes arrive in large numbers all at once from emergency rooms and critical care units. It is essential that you conduct the tests efficiently and accurately. It is also vital that you record your results accurately and clearly. You will be trained in **routine procedure**s (repetitive tasks with very standard sets of directions) in each of the five major areas of a clinical laboratory. These are

- bacteriology
- blood bank (Figure 1)
- clinical chemistry
- hematology
- urinalysis

The overnight shift consists of three or four other CLTs in addition to two phlebotomists (responsible for drawing blood samples) and a clerk.

Your Role in Solving the Problem

Your job is to learn how to process specimens in each of the five main laboratory areas for the day shift and perform a limited number of routine and **stat procedures** (those that must be completed immediately). When you finish your training, you will be given a final problem—to work the overnight shift. Your training evaluation will be based on the decisions you make as you handle the specimens you receive and the accuracy of your test results.

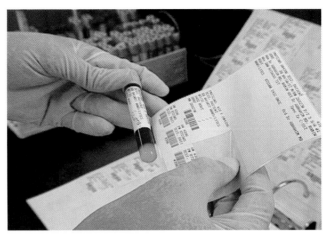

In the clinical laboratory, technical accuracy is not only important, it can be critical to a patient's life. As a CLT in training, you will need to become skilled using controls and standards to ensure that the results you report to your supervisor are accurate and precise.

You will learn to organize work schedules, prioritizing specimens as they enter your work area.

You will learn to keep accurate records, properly labeling all specimens, and placing them in appropriate containers for the tests ordered. You will learn to examine incoming specimens for quality and quantity. Specimens that do not meet the guidelines for testing must be rejected.

During the course of this module, you will learn

- the proper handling of patient specimens
- processing of specimens
- basic laboratory and safety procedures, and
- handling of reports.

overview of the workplace

Fact Sheet on Medical Laboratory Technology

Medical doctors, nurses, and nurse practitioners are the medical personnel that most patients recognize easily. However, it is the large support staff in hospitals and other settings that enables doctors and nurses to do their jobs well. The staff includes many different types of professionals, all of whom have varying degrees of training. They include

- biomedical technicians (responsible for equipment maintenance and repair)
- phlebotomists (responsible for drawing blood and assisting in getting specimens to the laboratory)
- respiratory therapists
- X-ray technicians
- clinical laboratory technicians
- medical technologists, registered nurses, and physicians of all types

Almost all hospital staff members are required to be licensed by a state or national licensing organization.

This module will focus on an important middle-level medical career—the clinical laboratory technician (CLT). The growing automation of many laboratory procedures is increasing the demand for CLTs and decreasing the demand for medical technologists (MTs), also known as clinical laboratory specialists (CLSs). The CLT is responsible for performing diagnostic tests on a variety of specimens, including blood and other body fluids. MTs serve as laboratory supervisors and also are responsible for performing tests that require the most specialized skills and knowledge. The MT reports to a laboratory manager and a pathologist.

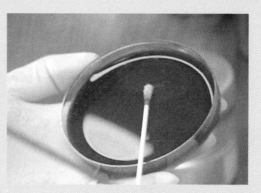

technician orientation

Training Objectives

By the end of the training program you, as a technician trainee, will be able to

- *set up* and *maintain* a laboratory notebook
- *follow* correct test procedures
- *identify* correctly labeled samples
- *follow* correct safety, health, and environmental procedures
- *define* key terms used in medical technology
- *describe* the role of blood and urine in the body
- *explain* why blood cross-matching is important
- *perform* a blood cross-matching test
- *perform* counts of red blood cells and white blood cells
- *perform* a glucose test for urine
- *transfer* a sample of bacteria to a slide
- *prepare* a heat-fixed slide
- *perform* a Gram-staining procedure
- *identify* whether a bacteria sample is Gram positive or negative
- *prioritize* sample test schedules for various situations

Developing Your Skills Base

The Medical Laboratory Technology module will help you develop laboratory and problem-solving skills that will be useful to you throughout this program and throughout your life. You will practice and refine these skills in the laboratory and through exercises on the CD-ROM. It will be up to you to monitor how well you are mastering a particular skill and to get help when you think you need it.

 *As you work, pay special attention to how you observe, measure, and test for the properties of various materials. Making* **quantitative and qualitative observations** *is a vital skill for laboratory technicians.*

Problem solving is also an essential technician's skill. Learning to use problem-solving strategies (science processes) will help you solve the drug development problem (and other science problems in the future) as efficiently and effectively as possible.

important concepts, processes, and skills

Science Concepts

When you finish this module, you should be able to *name/describe* and, where appropriate, *give examples of*

Multimedia Link

 Information on each of these Science Concepts can be found in the Reference section of the Medical Technology Laboratory CD-ROM.

medical technology terms

red blood cells

white blood cells

other blood components

correct ranges for red and white blood cell counts

cross-matching procedures

blood groups

urine components

Gram-staining reagents

Gram-staining procedures

controls in various procedures

types of diabetes

the role of glucose test in diagnosing diabetes

Science Processes

When you finish this module, you should be able to

define problems to be investigated

prioritize problems to be investigated

generate and *test* questions and hypotheses

conduct a valid analytical test

gather and *record* data

organize data into charts and graphs

analyze data

arrive at conclusions based on data

communicate results

relate test results to their implications for the body

Laboratory Skills

When you finish this module, you should be able to

prepare and *store* samples safely

follow directions for a clinical test (red and white blood cell counts, blood cross-matching, urine glucose test, blood electrolyte test)

light a burner

transfer a bacteria sample to a slide

prepare a heat-fixed bacteria slide

maintain sterile laboratory conditions

perform a correct Gram-staining procedure

use an oil immersion lens correctly on a microscope

follow appropriate safety procedures

Workplace Skills

When you finish this module, you should be able to

keep an accurate log of all laboratory work

read, understand, and *follow* written directions

follow safety/health/environmental guidelines

anticipate the work load and meet deadlines

prioritize tests and carry these out efficiently and accurately

access computers and other resources as necessary to find information

work cooperatively with team members

develop and *deliver* a clear and accurate presentation of results

Step 1 Setting Up and Maintaining a Laboratory Notebook

Materials

- bound notebook
- pen

Legal Issues

Laboratory notebooks are used to support the work of technicians in all types of laboratories. A technician's laboratory notebook must be updated on a daily basis. To be legally acceptable, each page in the notebook must be numbered, dated, signed by the technician, and witnessed by another person. The laboratory notebook can also serve as a record of the safety procedures that were followed for a test or an investigation.

Because others may need to read your results, it is very important that your laboratory notebook be neat, legible, and accurate. Dating pages makes it is easy for someone else to follow the sequence of all your experiments and tests. In addition, you should record any unique or new event with a notation of the exact time the event took place.

In the medical technology laboratory, there is typically a log for the particular type of test that is being run (Figure 1). Information about such things as the sample, test results, identity of the techni-

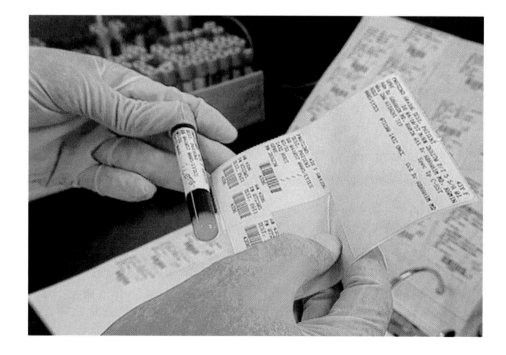

Figure 1
Checking the identity of a sample.

cian running the test, date, and time must all be faithfully recorded. As a back-up to these procedures, you will keep a personal notebook throughout the medical technology module. It is important that you record your procedures, test results, date, time, sample numbers, and any other pertinent information clearly and accurately in your notebook.

Procedure (for each student)

Begin by entering your identification information on the inside of the front cover. As a technician in a large laboratory, you will need to include information identifying your location, your group, and your workstation.

Each team member should keep his/her own laboratory notebook. It should contain a record of all

- sample identification
- predictions
- tests and procedures
- data
- observations
- calculations
- notes
- conclusions

Technicians follow standard procedures for tests. The standard test used to measure a particular substance *must* be recorded in the laboratory notebook. You may also be required to include a description of each complete experiment and a summary of all tasks completed by your team.

⏻ *At the end of each class period, remember to have your notebook witnessed and signed off by another member of your team.*

Handling Data

Data must *always* be labeled and described. Be sure to record

- the identity of every sample that is tested
- the time of day the test was run
- environmental conditions (temperature, humidity) at the time the test was run
- your observations about the samples

Including observations about samples as part of your data is an essential part of a medical technician's notebook entries. Observations about medical tests could include red and white blood cell counts, cross-matching information, urine glucose amounts, and many other things.

💧 *What you write about one sample distinguishes it from any other.*

💧 *A table is always a good way to present data, as it is easy to review later on.*

As you perform the various steps in your training program, you should return to this section. Use the information in Step 1 (Setting Up and Maintaining a Laboratory Notebook) as a checklist to plan and evaluate the accuracy of your record keeping.

Connecting to the Problem

Although a procedure and its results are very familiar to a technician while he/she is running a test, it can be hard to remember later just what a sample looked like or what results the test produced. This is why a well-kept laboratory notebook (and log, in the case of medical technology) is necessary. It serves as a complete history of every sample and test result. No information is unimportant. To prepare for your final evaluation as a clinical laboratory technician on the overnight shift, you will need a written record about each activity.

💧 *Only clear, accurate records of data and procedures will do for your laboratory notebook.*

Multimedia Link

 In addition to your laboratory notebook, you will have an electronic notebook on the CD-ROM. You can copy text from the CD into this notebook, edit it, print it, and save it to a floppy disk.

As you begin your training as a CLT, you will learn about some of the automated tests performed in the clinical setting and what the test results mean to both doctor and patient. Then you will compare the automated tests to the traditional tests on which the automated tests are based.

Step 2 The Impact of Automation in the Clinical Laboratory

Activity Overview

Once performed "by hand" by skilled technicians, clinical laboratory testing now relies on a series of instruments that not only do the analyses but also provide printouts of results. To understand the impact of automation, you will investigate how certain tests were conducted before automation in the laboratory became common.

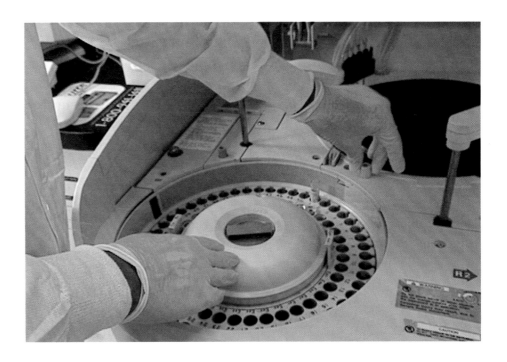

You will consider the advantages and disadvantages for automating these laboratory tests.

Resources (for the whole class)

- medical and clinical laboratory technicians as advisors (if available)
- medical library
- medical reference materials in community college or high school library
- Web access

Safety

There are no special safety requirements for this research step of the training.

Procedure (for 3 or 4 team members)

1. Table 1 suggests a way to organize the information you gather about automation in the clinical laboratory. Share the research task with other teams of CLTs in your class. Resources that you may want to use can include medical laboratory personnel from local laboratories and hospitals as well as medical, nursing school, community college, or university libraries.

2. This is a list of procedures and tests that you and your fellow CLTs can research. Your teacher may have others to add to the list.

- blood bank
- type and cross-match blood

Procedure/ Test	Original Instruments (before automation)	Current Automation	Advantages of Automation	Disadvantages of Automation	Source of Information
Blood bank					
Type and cross match blood			*FOR REFERENCE ONLY*		
Red blood cell count					
Others on list					

Table 1
Laboratory Automation

- red blood cell count
- white blood cell count
- white blood cell differential
- blood glucose
- physical and chemical analysis of urine
- hematology/urinalysis
- liver and cardiac enzymes
- general bacteriology
- specimen processing
- aseptic technique
- Gram stain
- preliminary cultures
- electrolytes (Na, K, and Cl) by ion selective electrode (ISE)
- clinical chemistry

Resources on the Internet

Note: URLs change. A search using the name of the organization will access the current Web location.

- American Diabetes Association
  www.diabetes.org
  Information about diabetes
- Healthlinks: Hematology
  www.hslib.washington.edu/clinical/hematology/html

Hematology reference works, cell images, journals and other resources

- Healthlinks: Microbiology
www.healthlinks.washington.edu/basic_sciences/micro.html
Microbiology resources
- American Association of Blood Banks
www.aabb.org
Bloodbanking and transfusion information
- National Institutes of Health

www.nih.gov
Resource for all aspects of health-related research including clinical trials
- The World Wide WebVirtual Library: Biosciences: Medicine at
www.ohsu.edu/cliniweb/wwwvl
An online database of medical information.

3. When you finish your research, share the information and resources you have collected with your fellow trainee CLTs. Together you may want to create a database of all your results.

4. When you have a complete set of information about various parts of the clinical laboratory that have been automated, make a list of all the instruments that a CLT would need to know how to operate. If possible, ask a medical technician to review the list and make suggestions.

5. If possible, visit a local clinical laboratory to see how automated instruments are used. Interview CLTs who have been in the profes-

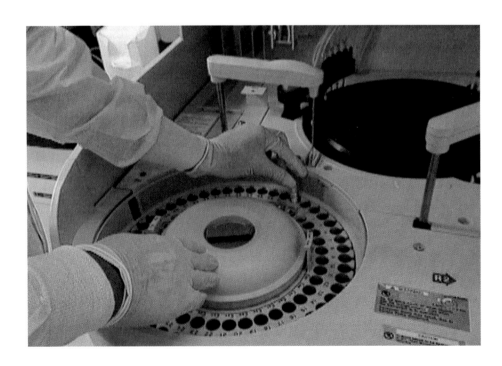

sion for a long time to find out how instruments and procedures have changed over the years.

Analyzing the Data

Review all of the information you have in your table. As a class, discuss the major advantages and disadvantages of automating the instrumentation used for clinical tests.

Arriving at Conclusions

Prioritize the advantages and disadvantages of automating test procedures on your list. For which tests do the advantages outweigh the disadvantages? **?** *note*📖📗

Step 3 Introduction to the Clinical Laboratory

The clinical laboratory is the place where doctors turn to get the information that patients cannot provide about their illnesses. Some procedures you will learn are common to all departments in the laboratory. Other procedures are specific to single departments.

In this laboratory, you will learn to

- use clinical terminology

- prioritize your work

- evaluate specimens as acceptable or unacceptable for testing

Part A Medical Laboratory Terminology

Language can be a barrier or a bridge. Many patients think that medical personnel communicate with one another in code. That is because many of the terms that medical professionals use are abbreviations and acronyms constructed from ordinary words used in everyday speech. Other terms come from Latin or Greek, the sources of many scientific words. This exercise will help you communicate clearly and concisely with your fellow CLT trainees using the language of the clinical laboratory workplace.

Materials

- laboratory notebook and pen

- resources used in Step 2 of your orientation

- reference tools on the CD-ROM
- Science Connection: The Blood, on page 15

Safety

There are no safety concerns for this exercise.

Procedure (for 3 or 4 team members)

1. Using all resources available to you, define the meaning of each of these terms as they are used in a clinical laboratory. Record your work in your laboratory notebook.

- anticoagulant
- ASAP
- CBC
- control
- EDTA
- electrolytes
- heparin
- plasma
- QA
- QC
- RBC
- routine
- rush
- serum
- sodium citrate
- standard
- stat
- type and cross
- UA
- WBC
- whole blood

2. Check your definitions with fellow technician trainees, so that you all have the same information about each term.

Part B Proper Specimens

All laboratory tests require specimen that are collected and treated in a very specific way. For instance, almost all hematology tests require whole blood with **EDTA** (ethyleneadiamine tetraacetate)

Blood may appear as a uniform red liquid, but it is actually a complex mixture of cells and chemicals. Red cells, a number of different kinds of white cells, and platelets occupy about 45% of the blood volume. They are suspended in a watery fluid called **plasma,** which consists of a variety of proteins and fats, dissolved salts (electrolytes), and glucose (blood sugar).

If you examine **red blood cells (RBCs)** (or, more technically, *erythrocytes*) under the microscope, they look like discs that were pinched in the middle. The biconcave shape of the disc exposes more surface area for oxygen exchange than a flat disc would. Mature red blood cells (Figure 2) have no cell nuclei—this cell structure is lost during early stages of the cells' development in the bone marrow.

RBCs get their red color from the approximately 250 million molecules of iron-containing hemoglobin found in each cell. Hemoglobin is responsible for transporting oxygen molecules to all parts of the body. Four iron atoms are present in each hemoglobin molecule; two of these form weak bonds with oxygen molecules (O_2) in the lungs. When these oxygen-laden hemoglobin molecules reach tissues in the body, the reaction reverses. Because conditions found in the neighborhood of active tissues are slightly acidic, the hemoglobin releases O_2 for use by the cells.

At the same time that oxygen is being delivered to body tissues, the plasma in the blood receives carbon dioxide (CO_2), the waste product given off by cells as they release energy from food molecules. Carried in the plasma, the CO_2 molecules remain in solution until reaching the capillaries of the lung. Then the CO_2 crosses thin membranes to the lungs and finally is released as a gas.

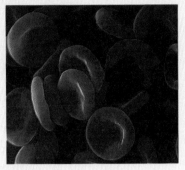

Figure 2

Red blood cells carry oxygen throughout the body.

Red blood cells are produced in the bone marrow. Clinical laboratory technicians count the red blood cells in a sample of blood to determine whether the RBC count falls within normal range for a patient's age and other body conditions. For instance, if the patient lives in high altitudes, the body may produce more red blood cells than normal. Variations from the normal range may have more serious causes, however. For instance, the body will have fewer red blood cells than normal if the patient is bleeding internally (a condition called *hemorrhagic anemia*) or has another type of anemia.

White blood cells (WBCs), or *leukocytes,* are also produced in the bone marrow. They not only defend the body from outside invaders, but they also screen out abnormal growing cells within the body, thus preventing cancer in the earliest stages. The tasks of white blood cells involve making antibodies, secreting chemical messengers, and engulfing foreign substances invading the tissues.

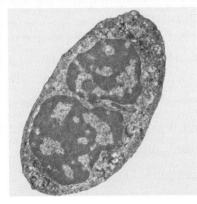
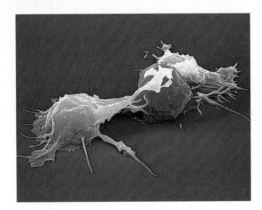

Figure 3

White blood cells. Active in the immune system, these cells have darkly staining nuclei that allow technicians to identify them.

White blood cells (Figure 3) are typically larger than red blood cells and come in a variety of types. Five basic kinds predominate:

- *Basophils* release chemicals like histamines to control blood vessel dilation.
- *Neutrophils* and *monocytes* are large cells that engulf and eat foreign material. *Eosinophils* engulf certain parasites and control the body's allergic reactions.
- *Lymphocytes* produce antibodies, specialized proteins that coat foreign materials.

As a group, basophils, neutrophils, and eosinophils are called *granulocytes*—a name describing the "grainy" appearance of these cells when stained and examined under the microscope (Figure 4). Granulocyte counts are elevated when a bacterial infection occurs, while lymphocytes tend to increase during a viral infection.

Certain disease-causing agents have adapted ways to use white blood cells as their entry points into the body. For example, when the tuberculosis bacterium enters the body, the germ is rapidly engulfed by WBCs. However, a waxy outer capsule prevents the germ from being digested. HIV viruses thrive in a type of WBC

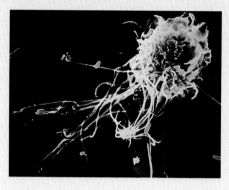

Figure 4
In phagocytosis, a white blood cell engulfs a foreign cell and destroys it.

found in body fluids outside the blood vessels—the "helper" T-cells. The HIV viruses replicate within the T-cells, ultimately destroying them. If the number of T-cells falls below critical limits, the body's ability to mount an effective immune response against other diseases is destroyed—becomes deficient. The result is AIDS, or acquired immune deficiency syndrome.

Blood platelets are fragments of cells pinched off from larger cells in the bone marrow. Containing a collection of cell catalysts called *enzymes,* platelets initiate reactions between certain plasma proteins to form blood clots. Access the Medical Technology CD-ROM for additional information on blood clotting.

added as an **anticoagulant,** a substance that keeps the blood in liquid form. Most blood chemical tests are done on cell-free plasma. Blood bank, hematology, and chemistry tests generally require blood specimens, while bacteriology tests are performed on a variety of tissue specimens. Urinalysis, as the name implies, deals almost exclusively with urine.

Collected blood specimens are delivered in tubes with a system of color-coded stoppers (Figure 5), which identify the type of additive that was present in the tube when the blood was collected:

- red tops: blood with no additive
- green tops: blood with the anticoagulant **heparin** (mainly used for research purposes)
- purple tops: blood with EDTA

Figure 5
Color-coded test tubes are
used to organize blood
specimens.

- blue tops: blood with added sodium oxylate, an anticoagulant and preservative
- tiger tops: blood to which has been added a black camouflage pattern; this cap indicates that a gel in the tube will keep the serum or plasma layer separate from the underlying sample of blood cells

Materials

- laboratory notebook
- pen
- samples of specimen containers (empty)
- reference materials

Safety

There are no safety concerns for this exercise.

Procedure (for 3 or 4 team members)

Study this list of specimen types:

- red tiger top tube
- plain red top tube
- blue top tube
- green top tube
- purple top tube
- urine sample

1. Using all your reference materials, identify each specimen container.

2. Indicate to which department (bacteriology, blood bank, chemistry, hematology, or urinalysis) a specimen in each container would be sent, and why. (*Note:* Some specimens can go to more than one department.)

3. Discuss your results with others in your class.

Analyzing the Data

Discuss these two questions with others in your class.

? *Why is it important to know the coding system for the specimen tubes?*

? *What are the possible consequences of placing the wrong cap on a specimen tube?*

Summarize the main points of this discussion in your laboratory notebook.

Part C Proper Patient Identification

All specimens must be accompanied by a **request slip** (see Figure 6a). This slip must contain the following information:

- the patient's name
- hospital identification number
- age
- tests requested
- time of order

Figure 6a
A sample laboratory request slip

Name:	John Doe
I.D. #:	55-5555
Sex:	M
Age:	35
Time requested:	10:30 A.M.
Date:	02/01/00
5250	CBC with diff
Priority:	Routine

```
Name:     John Doe

I.D. #:   555-55-5555

Date:     2/1/01

Code:     1100
```

The slip must also have the priority of the test clearly marked and must be labeled correctly (Figure 6b).

Information on the specimen label must include

- the patient's name and hospital identification number

- the date and time of collection

- the initials of the person who collected the specimen

Multimedia Links

The Technician Tasks section in the CD-ROM shows samples of specimen labels being used to identify the patients from whom the samples have come and to show the results of testing. The video also shows how careful identification procedures are followed.

If any information does not match on both the specimen and the order slip, the specimen is recollected. For example, if the same patient's name appears on the two slips but there are different i.d. numbers, the specimen would be rejected. Why?

You will be provided with a series of specimens and order slips. Your job is to deliver the specimens to the correct departments. You must reject any specimens that are incorrect for that department or any that are mislabeled.

Materials

- laboratory notebook
- pen
- reference materials
- order slips
- labeled specimen containers

Safety

There are no safety concerns for this exercise.

Procedure (for 3 or 4 team members)

1. Set up a table in your notebook to serve as a log sheet, with columns for specimen number, patient name, patient ID, and department to which the specimen was routed. See Table 2 for an example of an appropriate form.

2. Obtain several numbered specimens and slips from your teacher.

3. Record the specimens on your log sheet, and note any rejected specimens with the word "recollect" in the department column.

4. Return rejected specimens to your teacher in exchange for the recollected ones.

5. Log in the new specimens.

Arriving at Conclusions

With your team members, discuss what you think would happen if an incorrect specimen was sent to each department. Consider both the case of a mislabeled specimen and a case of an improper specimen. Discuss the following questions with your team, and summarize your discussion in your notebook. note

? *What might cause a specimen to be mislabeled?*

? *Why might a technician collect the wrong specimen for a test?*

? *How does the labeling procedure seek to prevent mistakes from happening?*

? *If a mistake is made, how would it show up under this labeling procedure?*

? *What actions would be needed in the case of an incorrectly labeled specimen?*

? *What actions would be needed in the case of an improper specimen?*

Table 2
Sample log sheet

Specimen Number	Patient Name	Patient ID	Department Receiving Specimen

Laboratory 1 Blood Banking

Laboratory Overview

The blood bank is one area of the laboratory where a small mistake can cost a patient's life.

The most important two tests you will be asked to do on the overnight shift are to determine the blood type of a specimen and perform a cross match with donor blood.

When you perform a **blood type test,** you are testing to determine whether the patient has type A, B, AB, or O blood. You are also testing for the presence or absence of the Rh group (the + or − in blood types).

When you perform a **cross match,** you are testing to ensure that the donor blood the patient may receive is compatible with the patient's blood, to avoid a transfusion reaction.

Step 1 Blood Typing

Materials

- safety goggles, apron, and disposable gloves
- artificial blood samples, provided by your teacher

Note: For health and safety reasons, you will be using artificial blood during your training.

- anti-A serum
- anti-B serum

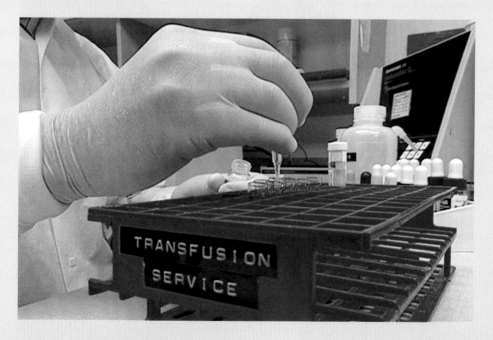

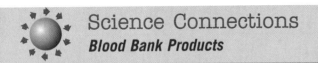

"Nurse, hang two units of O neg. Stat!"

Anyone who has watched a television show or movie about a hospital has probably heard that line. But what is a "unit?" What does "O neg" mean? What is "Stat"?

The last question is the easiest. **Stat** is a term used by medical professionals that means, "I want it done RIGHT NOW." "O neg" refers to blood that is type O and Rh negative. (See the Science Connection on page 27 for information about blood types.) A "unit" is one standard sized package of blood product, taken from one donor, at one donation.

In the early days of transfusions, a doctor would place a needle in the donor's arm and attach the needle with a tube to another needle that was then placed in the recipient's arm. Blood flowed directly from the donor, ensuring that the recipient received a fresh, uncontaminated transfusion. There was no way to collect blood and save it for use later. Also, little attention was given to matching blood types. As a result, the success rate of transfusions was low, although no one understood why. Doctors now know that the consequences of transfusing mismatched blood can be fatal.

With the development of anticoagulants and preservatives, whole blood can now be collected from donors, stored, and used at a later date. You are probably familiar with "blood drives" that take place in your area. When people respond to a blood drive, they donate their blood for medical use by others. The collected blood is stored in a *blood bank*. The normal adult human body contains about four to six liters of blood in circulation. Until recently, one pint of blood has been the acceptable quantity for a person to give at any one donor session. Today, since under controlled conditions the body normally replaces limited blood loss quickly, donors are permitted to give as much as two pints at a single session.

To store blood over long periods of time, cells are separated from the plasma, and both are properly stored at low temperatures. It is now possible to isolate several components from whole blood; each component is useful for the treating disease. As a result, blood collected from a donor may be separated into

- packed red blood cells
- platelets
- plasma
- clotting factors, which are proteins extracted

from plasma and used to treat hemophilia.

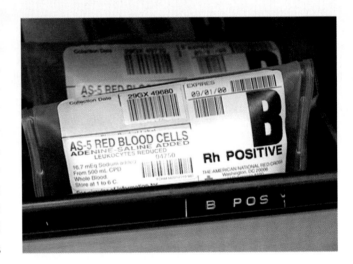

- anti-Rh serum
- testing card or index cards on which are drawn three circles labeled A , B, and Rh
- clean toothpicks or plastic stirrers

Safety

- Protective gloves, safety goggles, and aprons must be worn at all times in a clinical laboratory to reduce accidental exposure to

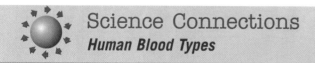

Although all human blood samples may have the same physical characteristics, the samples may differ in an important chemical property from one person to another. There are several distinct **blood types—A, B, and O—** based on the presence or absence of certain marker molecules on the membranes of red blood cells. These surface markers are called **antigens.** The ABO blood group system is one of the main antigen systems in your body. All of the red blood cells in your body are marked in the same way. They either have the A antigen only (type A blood), the B antigen only (type B blood), both A and B antigens (type AB blood), or neither A nor B antigens (type O blood).

If a patient receives blood carrying an antigen that does not match the antigen already found on the patient's blood, the body's immune system mounts a swift defense. Soon, antibodies produced in the patient's blood destroy the "foreign" donor cells—a condition called *hemolytic transfusion reaction.* The effect of this incompatibility is severe. The first symptom is pain at the site of the transfusion, soon followed by chest and back pain, chills, and fever. In extreme cases, the reaction may result in kidney failure and death.

A second antigen system found on blood cells, independent of the AB system, is the Rh system, which is named for its initial discovery in rhesus monkeys. If a person's red blood cells carry the Rh antigen, the person is Rh positive, or Rh+. A person lacking the antigen is Rh negative, or Rh−. If a person is Rh−, the person's blood will have antibodies for the Rh antigen only

if the person has been exposed to the Rh+ antigen through transfusion, bleeding across the placenta during pregnancy, or some other exchange of blood. The build-up of antibodies for Rh antigen takes time. When an Rh− individual is first exposed to Rh+ blood, there is no immediate reaction. But this first exposure sensitizes the patient to the Rh antigen. The risk of a hemolytic reaction occurs on the second exposure.

An Rh− woman who gives birth to an Rh+ baby can be sensitized (develop antibodies) if there is even a small exchange of blood between the mother and the developing fetus, perhaps through a tear in the placenta. The brief exposure sensitizes the mother to the Rh antigen, almost as if she has been vaccinated. Neither she nor her first child are at risk. But if she becomes pregnant with a second Rh+ baby, her anti-Rh antibodies are ready to cross the placenta and attack the baby's red blood cells, which will cause massive cell destruction, a condition called *hemolytic disease of the newborn.* At best, the baby will require treatment in the intensive care nursery and possibly a transfusion. At worst, the baby could require all of its blood to be replaced, or the baby may die.

To prevent hemolytic disease of the newborn, doctors make sure that they are informed about the blood types of both mothers and babies and treat at risk pregnant patients with injections of serum containing anti-Rh antibodies. The procedure "fools" the body into accepting that the antibodies are already being produced. The production of new antibodies is halted.

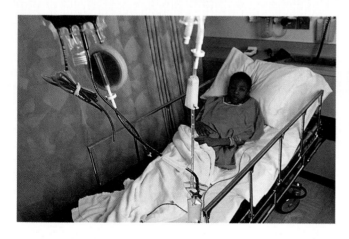

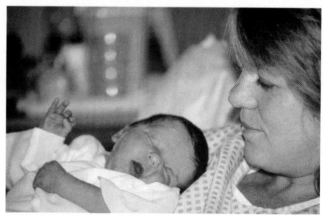

infection by blood-borne diseases. Although you are using artificial blood in this laboratory, you will comply with this safety procedure.

Procedure (for pairs of students)

1. Put on gloves, safety glasses, and an apron.
2. Place all anti sera within easy reach.
3. Place a large drop of artificial blood in each circle on the card.
4. Add one drop of anti-A to circle A. Stir with a toothpick.
5. Add one drop of anti-B to circle B. Stir with a new toothpick.
6. Add one drop of anti-Rh to circle Rh. Stir with a new toothpick.
7. Observe the blood for agglutination (clumping) as you stir the serum and blood.
8. Make a labeled sketch in your notebook displaying the results of the three tests.
9. Use Table 3 to interpret your results:

Multimedia Links

The Technician Tasks section in the CD-ROM and the video both show blood type testing taking place in a medical laboratory. View these to compare your work with the work of a CLT.

Blood Type	Anti-A	Anti-B	Anti-Rh
A+	●	○	●
A−	●	○	○
B+	○	●	●
B−	○	●	○
AB+	●	●	●
AB−	●	●	○
O+	○	○	●
O−	○	○	○

Table 3
Blood Typing

Analyzing the Data

? *What is the blood type of your specimen?*

? *Which antigens—A, B, and Rh—were present in your specimen? Explain your answer.*

? *Consult the American Association of Blood Banks (AABB) Technical Manual to determine the frequency with which the different blood types occur in the general population. Would the blood type of your specimen be considered a rare blood type?*

Arriving at Conclusions

For each of the blood types in the blood typing table, which antibodies would be present in the blood of the patient from whom the sample was drawn—anti-A, anti-B, anti-Rh? Explain your answers.

While every attempt is made before a transfusion takes place to match the blood types of the donor and the recipient, certain exceptions to this rule are possible in an emergency situation.

From the following list, predict which procedure would put the recipient at the LEAST risk. Explain your choice in your notebook.

note

1. Type A+ donates to Type B+

2. Type AB− donates to type A+

3. Type O− donates to type AB+

4. Type AB− donates to type O+

? *Which blood type listed in the blood typing table is safest for "universal" donations? That is, which blood type could be safely given to any patient?*

? *Which blood type would a patient need to make him or her a "universal recipient," able to receive donations from all other blood types? Explain your answer.*

Connecting to the Problem

Suppose that you are medical technologist working the night shift in a clinical blood bank lab. If all of these work orders came in at the same time, how would you prioritize them? **?**

1. A woman who is in labor, admitted to the hospital to have a baby normally.

2. A patient who is going to have surgery the following morning.

3. A patient who has received a gunshot wound to the belly.

4. A patient in the emergency room with fever and chills.

5. A woman scheduled to have a baby by Caesarian section in the morning.

Explain your decisions in your laboratory notebook. *note*📖

Step 2   The Cross Match

When a patient is going to receive a transfusion, all units of blood must be tested with the patient's serum to make sure that the blood groups are compatible. Even though the type of each unit is clearly marked on its container, a technician must verify the information and cross-match the two blood types to be sure that there will not be any adverse reaction when the blood is transfused. Any human error at any point in the donation process or during the cross match could result in the death of a patient.

Materials

- protective gloves, goggles, and an apron for each person
- artificial donor blood specimen
- artificial patient blood specimen
- anti sera from Part 1
- blood typing cards
- toothpicks or plastic stirrers
- 10% chlorine bleach solution

Safety

- Protective gloves, aprons, and safety goggles must be worn at all times in the clinical laboratory to reduce accidental infection with blood-borne diseases.

- At the end of the activity, all surfaces must be cleaned with a solution of 10% chlorine bleach to kill any biological contaminants.

Procedure (for 3 or 4 team members)

1. Obtain two specimens of simulated blood as assigned by your teacher. One will be labeled "donor," the other "patient."

2. Type the donor blood and patient blood as in Step A. Record the results.

3. Mix a drop of donor blood and a drop of patient blood, stir with a new toothpick, and observe for agglutination (clumping).

4. If no agglutination occurs, the two specimens are compatible, and the patient can receive the blood. If agglutination occurs, the specimens are incompatible, and the donor blood is unsuitable for the patient.

Analyzing the Data

After discussing these questions with members of your team, record your answers in your laboratory notebook.

? *What would you do if you found a unit of blood that was mislabeled after you tested it? Talk to a local CLT to find out what he or she would do.*

? *What causes the agglutination reaction to occur in incompatible samples? Use an example of an incompatible donor-recipient match, and tell what is present in each specimen to cause agglutination to take place.*

Arriving at Conclusions

Discuss this question with the members of your team: What do you think would be the best course of action if an accident victim came into the emergency room needing an immediate transfusion, and there was no time for a cross match? Justify your recommended course of action. Record the main discussion points in your notebook. **?** *note*📖

Connecting to the Problem

Remember that you are on the overnight shift. Under what circumstances would you give blood bank work priority over all other departments? Justify your opinion. Record your answer for use in the final laboratory. **?** *note*📖

Laboratory 2 Blood Cells

Part A Red Blood Cell Count

Laboratory Overview

How can we tell just how many red blood cells there are in our blood? The answer is easier than it seems; we simply count them. First, a small sample of blood is diluted. Next, the diluted blood is placed in a chamber on a special microscope slide that is called a **hemacytometer** (Figure 7), which is used to provide a standard reference for counting the cells. After the cells are counted, a mathematical calculation gives the number of red blood cells per milliliter of blood.

Materials

- RBC diluting pipette (red bead in mixing chamber) or a unopette, if available
- appropriate suction device for RBC pipette
- isotonic saline solution
- artificial blood
- hemacytometer and cover slip
- alcohol pads
- lint-free laboratory wipes
- hand counter
- microscope

Figure 7
A hemacytometer

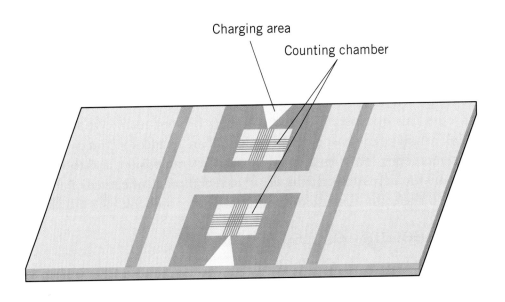

Charging area

Counting chamber

Safety

- Wear safety goggles, plastic gloves, and a laboratory apron at all times in the clinical laboratory to reduce accidental infection with blood-borne diseases.
- Be careful not to crack the hemacytometer cover slip when cleaning it.
- Do not touch or spill any of the liquids.
- In case of accidental exposure, wash the spill with soap and sponge, and clean skin with hand soap and towels.
- Wash hands thoroughly with soap on completion of this laboratory experiment.
- At the end of the activity, all surfaces must be cleaned with a solution of 10% chlorine bleach to kill any biological contaminants.
- Dispose of all materials as directed by your teacher.

Procedure (for 3 or 4 team members)

1. Clean the hemacytometer and cover slip with an alcohol pad, and dry with a lab wipe. Make sure the RBC pipette (Figure 8) is clean.

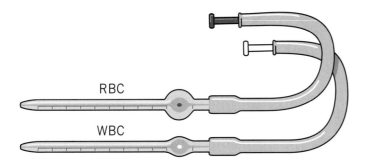

Figure 8
A blood-diluting pipette

2. Attach the pipette to the suction device. *Note:* Never use suction by mouth.

3. Draw blood into the pipette to the 0.5 mark. If you draw a little too much blood, you can remove the excess by tapping the pipette lightly on a nonabsorbent material. If the blood is more than 2 mm above the 0.5 mark, clean the pipette and start again.

4. Clean the outside of the pipette *without touching the opening.*

5. While gently rotating the pipette, quickly and carefully draw normal saline to the 101 mark. Do not overshoot the mark or allow bubbles to form in the pipette.

6. Clean the outside of the pipette *without touching the opening*.

7. Roll the suction tubing off of the pipette, being careful not to expel any of the sample.

8. Hold the pipette and rotate it in a figure eight pattern for 2 to 3 minutes (Figure 9). If using a unopette, follow the manufacturer's directions for filling, mixing, and dispensing.

Figure 9
Rotate to mix blood and dilution fluid in the pipette's distillation chamber

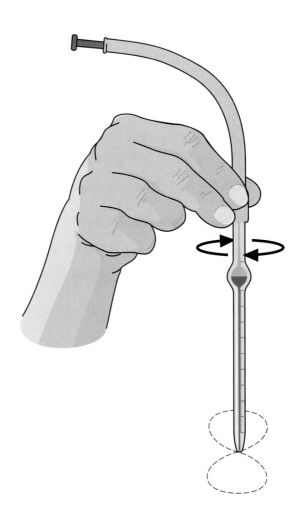

9. Expel 4 to 5 drops from the pipette to remove any unmixed saline.

10. Place the cover slip on the hemacytometer, and charge both chambers by touching the tip of the pipette to the cover slip edge where it meets the chamber floor (Figure 10). If fluid overflows into the moat, you will need to clean the hemacytometer and refill the chambers.

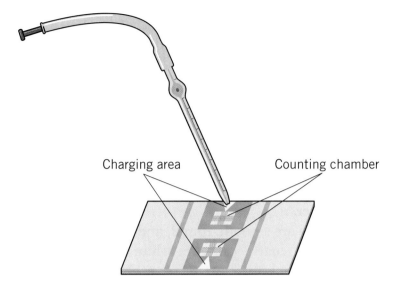

Figure 10
Charging the hemacytometer chambers

Charging area

Counting chamber

11. Lay the pipette in a horizontal position, and save the contents until you are sure you will not need another sample.

12. Mount the hemacytometer on the microscope, and adjust the light setting to low. Let the hemacytometer rest for 1 to 2 minutes to allow the diluted sample to settle (Figure 11).

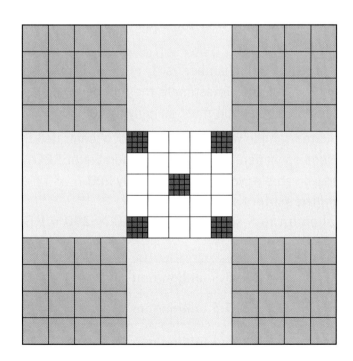

Figure 11
The counting chamber of the hemacytometer squares. Step 13 tells you how to count the cells in your RBC sample.

13. Counting RBCs:

   a. Scan the slide under the 10x objective to ensure that the cells are evenly distributed.

   b. With the 40x objective (or "high-dry" objective) in place, count all of the RBCs in the five small ruled RBC squares within the central counting chamber. Start with the upper left corner square, then the upper right, the center, then lower left, and, finally, the lower right. Each of these squares is further subdivided into 16 smaller squares to make your counting easier.

   c. Red blood cells are smooth, round, and somewhat doughnut-shaped. Larger, coarse cells are white blood cells. Do *not* count WBCs. Click the counter for each RBC you see.

   d. As you count the cells within one of the RBC squares, follow this rule to count the cells at the boundaries: Count all of the cells that touch the upper and left boundaries of the sub-square, but do not count the cells touching the bottom or right boundaries in the sub-square.

   e. Record the number of RBCs counted in all five ruled RBC squares.

Calculate the number of RBCs per cubic millimeter of the original sample using the following method:

- There are a total of 25 squares within the complete grid. You counted cells in only 5 of the squares, so you must first **multiply your count total by 5.**

- The entire central chamber is 1 mm $\times$ 1 mm $\times$ 0.1 mm = 0.1 mm^3 in volume. To estimate the number of cells in a mm^3 of the diluted sample, you must **multiply by 10.**

- The diluted sample was 200 times less concentrated than the original blood specimen. To estimate the number of RBCs in a mm^3 of the drawn sample, you must **multiply by 200.**

In summary, the patient's RBC count is

(Total count in 5 squares) $\times$ 5 $\times$ 10 $\times$ 200 = RBCs/mm^3

Table 4
RBC Averages

Newborn:	4.4–5.8 million/mm^3
Infant/Child:	3.8–5.5 million/mm^3
Adult Female:	4.2–5.4 million/mm^3
Adult Male:	4.7–6.1 million/mm^3

Analyzing the Data

Table 4 gives the normal ranges for RBC counts.

Consult a medical textbook or a human anatomy and physiology textbook to find the answers to the following questions. Record your answers in your laboratory notebook.

1. What could account for the differences in the ranges in the different groups? ❓

2. Describe at least two different conditions that would cause the red blood count to go down. *note*📖

3. Describe at least two different conditions that could cause the red blood count to go up. *note*📖

Connecting to the Problem

Some conditions that would cause the red blood cell count to change will also affect results in blood chemistry and possibly may require the administration of blood products from the blood bank.

❓ *How would severe dehydration be expected to change the RBC count in a patient? Would dehydration also affect the electrolyte concentration? Explain.*

❓ *Would you expect that a doctor would order a blood transfusion for a severely dehydrated patient? Explain your reasoning.*

❓ *Do you think a rapid drop in RBC count could be an indication that a patient needs a transfusion? Explain your answer.*

Part B White Blood Cell Count

Laboratory Overview

The procedure for counting white blood cells is similar to that for counting red blood cells. WBCs differ from RBCs in size, shape, color, and contents. Before beginning Part B, examine the photographs on pages 38–39 showing the unique properties of each.

Materials

- WBC diluting pipette (white bead in mixing chamber) or a unopette, if available
- appropriate suction device for WBC pipette
- 2% acetic acid by volume
- hemacytometer and cover slip
- artificial blood specimen

- alcohol pads
- lint-free laboratory wipes
- hand counter
- microscope

Safety

Wear safety goggles, plastic gloves, and a laboratory apron at all times in the clinical laboratory to reduce accidental infection with blood-borne diseases.

- Be careful not to crack the hemacytometer cover slip when cleaning it.
- Do not touch or spill any of the liquids.
- In case of accidental exposure, wash the spill with soap and sponge, and clean skin with hand soap and towels.
- Wash hands thoroughly with soap on completion of this laboratory experiment.
- At the end of the activity, all surfaces must be cleaned with a solution of 10% chlorine bleach to kill any biological contaminants.
- Dispose of all materials as directed by your teacher.

Procedure (for 3 or 4 team members)

1. Clean the hemacytometer and cover slip with an alcohol pad and dry with a lab wipe. Make sure the WBC pipette is clean.

2. Attach the pipette to the suction device. *Note:* Never use suction by mouth.

3. Draw blood into the pipette to the 0.5 mark. If you draw a little too much blood, you can remove the excess by tapping the pipette lightly on a nonabsorbent surface. If the blood is more than 2 mm above the 0.5 mark, clean the pipette and start again.

4. Clean the outside of the pipette without touching the opening.

5. While gently rotating the pipette, quickly and carefully draw 2% acetic acid to the 11 mark. (The acetic acid causes the RBCs in the sample to *lyse*, or break apart, so that all that remains for counting are the larger WBCs.) Do not overshoot the mark or allow bubbles to form in the pipette.

6. Clean the outside of the pipette without touching the opening.

7. Roll the suction tubing off of the pipette, being careful not to expel any of the sample.

8. Hold the pipette and rotate it in a figure eight pattern (see Figure 9, page 30) for 2 to 3 minutes. (If using a unopette, follow the manufacturer's directions for filling, mixing, and dispensing.) Expel 4 to 5 drops from the pipette to remove any unmixed solution in the stem.

9. Place the cover slip on the hemacytometer, and charge both chambers by touching the tip of the pipette to cover slip edge where it meets the chamber floor. If fluid overflows into the moat, you will need to clean the hemacytometer and refill the chambers.

10. Lay the pipette in a horizontal position and save the fluid until you are sure you will not need another sample.

11. Mount the hemacytometer on the microscope, and adjust the light to a low setting. Let it rest for 1 to 2 minutes to allow the diluted sample to settle.

Counting WBCs:
a. Scan the entire counting chamber under the 10x objective to ensure that cells are evenly distributed.
b. Using the 10x objective, count all of the WBCs in the upper left square within the counting chamber (the 1 mm × 1 mm square should take up approximately the entire field of view at this magnification.)
c. Apply this rule for counting cells at the boundaries: Count all of the cells that touch the upper and left boundaries of the square, but do not count the cells touching the bottom or right boundaries in the square. Click the counter with each WBC you see.
d. Record the number of WBCs counted, and repeat the counting technique for 3 more squares: upper right, lower left, lower right. (These are the squares that are shaded purple in Figure 11 on page 31.)
e. Add together your 4 counts, and record the total number of WBCs in the 4 counting squares.

Estimate the total number of WBCs in 1 mm³ of the undiluted blood sample:

- You counted all of the WBCs present in 4 squares. To find the average number of WBCs in a 1 mm × 1 mm square, you must **divide the total number of WBCs by 4.**

- The volume of one of the squares is 1 mm × 1 mm × 0.1 mm = 0.1 mm³. To estimate the number of WBCs in 1 mm³ of diluted blood, you must **multiply your average count by 10.**

- Your diluted sample is 20 times less concentrated than the original blood sample. To estimate the number of cells in 1 mm³ of the drawn sample, you must also **multiply by 20.**

In summary:

$$\frac{\text{Total WBCs in 4 squares}}{4} \times 10 \times 20 = \text{WBCs} / \text{mm}^3$$

Table 5
Normal WBC Range

Newborn:	9,000–30,000/mm^3
1-week-old infant:	5,000–21,000/mm^3
1 month old:	6,000–17,500/mm^3
2 year old:	6,200–17,000/mm^3
Adult:	4,800–10,800/mm^3

Analyzing the Data

Table 5 shows normal ranges for WBC count.

Using the medical reference books you have, answer the following questions in your laboratory notebook:

❓ *What could account for the differences in the ranges in the different groups?*

❓ *Describe at least two medical conditions that could cause the white blood count to go down.*

❓ *Describe at least two medical conditions that could cause the white blood count to go up.*

Connecting to the Problem

Suppose that a patient is experiencing severe lower abdominal pain, localized on the right side. Why would a doctor order a WBC count? What might be the consequence of assigning the order a low priority and delaying the test results? ❓ *note*📖

Part C Differential Cell Count

Laboratory Overview

While knowing the total number cells in a patient's blood is helpful in diagnosing many conditions, sometimes it is necessary to go a step further. Technicians can determine the proportions of the different types of cells. This blood analysis is called a **differential cell count.**

The differential WBC count is the key to diagnosis and treatment of anemias and leukemias. Other diseases and conditions—severe allergic reactions, typhoid fever, appendicitis, infectious mononucleosis and others—can also cause changes in the appearance and distribution of white blood cells. On page 38 you can see drawings of the five types of white blood cells.

For this test, you will view a prepared slide called a *blood smear.* Although both RBCs and WBCs will appear on the slide, the WBCs will be easy to spot. A stain called Wright's stain has been applied to the blood smear. It will change the color of the WBC nuclei—

stain them—so that they appear purple. The RBCs, which have no nuclei, will appear in the background as much smaller cells with the characteristic donut shape. Using an oil immersion objective, you will count WBCs in two fields, noting how many of each type you find. While you are counting WBCs, you will also note the morphology (shape) of the RBCs. Abnormal RBC morphology is key to diagnosing conditions like sickle cell anemia, an inherited blood disorder characterized by red blood cells that have a characteristic sickle shape.

Materials

- blood smears stained with Wright's stain, taken from both normal and abnormal patients
- microscope with oil immersion objective
- immersion oil
- paper and pen, or differential counter
- protective gloves, safety goggles, and an apron
- a color atlas of blood cells, such as *The Morphology of Human Blood Cells* by Diggs, Sturm, and Bell

Safety

- Wear safety goggles, plastic gloves, and a laboratory apron at all times in the clinical laboratory. Although the slides have been fixed with methanol, treat all blood smears as though they are potentially infectious. Wear gloves during the procedure.
- Use proper microscopy technique.
- Handle the slides carefully to avoid cracking or breaking.
- If there is breakage, do not directly handle any broken glass.
- Clean the work surface and return materials as directed by your teacher.
- Wash hands thoroughly with soap on completion of this laboratory experiment.

Procedure (for pairs of students)

1. Place the slide on the microscope stage, and observe under low power. Locate a part of the smear where the cells appear evenly spread out.

2. Switch to high-dry power, focus; then swing the objectives away so that they are clear of the surface.

3. Place a drop of immersion oil on the surface of the slide over the area you have selected (Figure 12). Swing the oil immersion

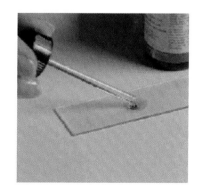

Figure 12
Placing immersion oil on a slide.

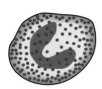

Basophil

Lymphocyte

Eosinophil

Neutrophil

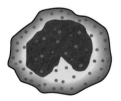

Monocyte

objective into place and carefully adjust the focus using the fine adjustment.

4. Take time to become familiar with the five basic WBC appearances. Be sure that you can recognize each type.

5. List the five WBC cell types (shown here) on a page in your laboratory notebook. Your partner should use this list as a tally sheet as you make your observations. Say the name of each WBC as you observe it. Continue to scan the field of view in a systematic back-and-forth pattern until you have counted all of the WBCs in that field.

6. Move the slide so that a nearby field is in view, and switch observing and tallying roles with your partner.

7. If unusual or abnormal cells are found, you must identify them and include them in the count.

8. After you have tallied the WBCs in two fields of view, look for and identify any abnormalities in the RBCs. Classify them, if present, as few, moderate, or many (1+, 2 +, 3+).

9. Determine the percent of the total represented by each of the WBC types on your slide. Do this by using the following formula:

$$\frac{\text{Number of WBC type}}{\text{Total number of WBCs counted}} \times 100 = \text{percent of type on slide}$$

Compare your data with the percents found in typical adults as shown in Table 6, page 39.

Analyzing the Data

Use the medical references you have available to answer these questions.

❓ *If any of your slides had results that were not within the expected ranges, what conditions or diseases are indicated?*

❓ *Did the RBCs look abnormal in any way? If so, what disease might cause them to appear this way?*

❓ *What type of WBC would you expect to find elevated in each of the conditions listed in the laboratory overview (p. 40)? What aspects of the condition accounts for the increase in cell numbers?*

Connecting to the Problem

An acute illness is one that appears suddenly and progresses rapidly either to death or a return to health. A chronic illness is exactly the opposite, a gradual decline in the patient's health. Again, refer to the list of conditions in the laboratory overview. Which would require a stat differential count to ensure rapid diagnosis for immediate treatment? For the choices you made, tell what the consequences of delayed treatment might be. ❓ *note* 📖

Type of WBC	Percentage
Neutrophils	40%–74%
Lymphocytes	32%–36%
Monocytes	4%–13%
Eosinophils	0–7%
Basophils	0–3%

Table 6
Percents of WBC in Typical Adults

Laboratory 3 Clinical Chemistry

Part A Electrolytes

Laboratory Overview

Electrolytes are ions found in the blood that help to maintain the body's osmotic balance. These ions are especially important for normal muscle and nerve functions. The most commonly measured electrolytes are sodium (Na^+), potassium (K^+), and chloride (Cl^-).

Materials

- Computer
- PASCO interface box
- Data Studio and Science Workshop software
- Ion Selective Electrodes (ISE) amplifier box (If possible, one for each probe)
- sodium ISE
- potassium ISE
- chloride ISE
- control solutions
- standard solutions
- unknown solutions

Safety

- Wear safety goggles and a laboratory apron throughout the procedure.

Procedure (for 3 or 4 team members)

1. Set up the probes, amplifiers, and interface boxes according to the directions supplied with the instruments. If possible, use a separate amplifier for each probe.

2. Start up the Science Workshop or Data Studio software.

3. Using millimoles per liter (mmol/L) as the calibration unit, calibrate the ISEs according to directions included with the probe.

Conversion Factors

Chloride	1000 ppm = 28 mmol/L
Potassium	1000 ppm = 13 mmol/L
Sodium	1000 ppm = 43 mmol/L

4. Test each ISE with its control solution and record your result. If the result is not within ±5% of the target value, repeat the calibration procedure.

5. Obtain the unknown samples from your teacher, and test each one.

6. Record the result in your notebook.

Analyzing the Data

Table 7 shows normal ranges for serum electrolytes:

1. Did all of your unknowns fall in the normal range?

2. Contact a clinical laboratory, and find out what values would alert a technician to contact the physician immediately. In other words, what are the "panic values" for these tests?

Table 7

Na^+	135—148 mmol/L
K^+	3.5—5.3 mmol/L
Cl^-	96—109 mmol/L

Arriving at Conclusions

Use the medical references you have available to answer these questions.

? *What conditions might be indicated by an increase in the concentration of each of the three electrolytes?*

? *What conditions might be indicated by a decrease in the concentration of each electrolyte?*

Connecting to the Problem

You have used ion selective electrodes (ISE) to determine the concentration of electrolytes in patient specimens. Based on your experiences with previous laboratory test results, what priorities should be assigned to running electrolyte tests? Under what conditions would they take priorities equal to or over other laboratory tests? **?** *note*

Part B Urinalysis

Laboratory Overview

Glucose is the main fuel source for your body. You get glucose from food, and your body uses it to provide energy for all of its functions.

In the clinical chemistry laboratory, most tests are run using sophisticated analytical instruments. To ensure accurate results, these instruments must be checked and calibrated periodically. To perform a calibration, the technician uses **standards**—usually aqueous solutions of known concentrations of particular substances (**analytes**) being tested such as sodium chloride, potassium chloride, or glucose. Usually, two or three standards are run to make a *calibration curve*, a graph that shows a reading—usually a flow of electrical current on a meter. A two-point calibration curve usually uses a low concentration or zero standard and a high concentration standard. A three-point curve uses a low, middle, and high concentration standard.

Other samples of known concentrations are called **controls.** In a control the sample is contained within the actual kinds of specimen material to be tested. In the clinical chemistry laboratory, controls are usually serum samples mixed with known concentrations of *analyte*. The technician looks for instrument readings that are ± 5% of the known value of the analyte in the control. If an instrument gives readings outside the ±5% range for the control, the instrument must be recalibrated with a standard to ensure its capacity to give accurate readings on actual specimens.

Specimens are the actual patient samples brought into the laboratory. The level of analyte in patient specimens is not known until the test is run on a calibrated instrument.

If you take in excess sugar, it is stored in your muscles and liver as glycogen, which is a starch. Eventually, the excess is converted to fat.

Diabetes may be diagnosed by measuring glucose in the blood itself or by testing samples of urine. Blood in the body passes through the kidney, which filters out harmful and excessive substances and excretes those substances as urine. Excessive glucose may be one of those excreted substances.

In this laboratory, you will test simulated urine specimens for glucose concentration, using prepackaged reagent sticks. A positive and a negative control solution is provided for you to pretest the reagent sticks.

Materials

- 1 bottle glucose reagent sticks
- color chart supplied with sticks
- labeled covered vials containing:

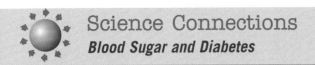

Science Connections
Blood Sugar and Diabetes

One of the most common disorders involving blood glucose levels is **diabetes,** a condition characterized by elevated glucose levels in the blood. Additional symptoms include fruity smelling urine, excessive thirst (polydipsia), excessive urination (polyuria), excessive hunger (polyphagia), and weight loss. A physician will diagnose diabetes in a patient based on a combination of these symptoms and elevated levels of blood glucose (Figure 13). The blood sample is taken under very specific conditions. The patient must fast—eat or drink nothing for about 12 hours before the test. Then the blood is tested. A fasting blood sugar level greater than 126 mg/dL in the presence of the other symptoms is indicative of the disease.

There are two major types of diabetes. Type I diabetes, which usually appears before a person is 30 years old, is characterized by a lack of insulin production. Type I is treated with insulin injections. Type II diabetes, which generally occurs after age 30, is characterized by a gradual decrease in insulin production. The standard treatment for Type II diabetes is prescription drugs that enhance insulin production. There are some new treatments for controlling diabetes. One of these is the insulin pump (Figure 14), a device that is implanted in the patient's body and that delivers a measured amount of insulin into the bloodstream as needed.

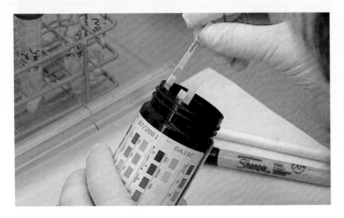

Figure 13
Diabetes test strips

Figure 14
An insulin pump is implanted in a diabetic person, then insulin is dispensed through the monitor.

- negative and positive controls
- simulated patient specimens
- protective goggles, plastic gloves, and laboratory apron for each student
- medical reference materials

Safety

- Wear safety goggles, plastic gloves, and a laboratory apron throughout the procedure.

- Although only simulated samples are being handled in this training laboratory, in an actual clinical laboratory, gloves are always worn when handling specimens of body fluids

Procedure (for 3 or 4 team members)

1. Dip a reagent stick into a positive control solution.

2. After exactly 30 seconds, read the concentration of glucose from the color chart.

3. If the color on the reagent stick is within the range of color values for the control, use the reagent stick to test patient specimens. If not, repeat the control test with a new stick.

4. Test and record a glucose concentration for each solution assigned to your team. Record your data on a table in your laboratory notebook.

Analyzing the Data

The normal concentration of glucose in blood is 65–115 mg/dL. Measurements outside of this range indicate disease conditions. For the specimen to test in the normal range, the color on the reagent stick you are using should match the color of the 100 mg/dL patch on the color chart (Figure 15).

? *Which of your patient specimens tested in the normal range? Which of your specimens tested outside the normal range?*

? *What is considered to be the panic number for urine glucose?*

Arriving at Conclusions

Consult your medical references to answer the following questions.

? *What conditions might result in hypoglycemia, a decreased blood sugar level?*

? *What conditions might result in hyperglycemia, an increased blood sugar level?*

Connecting to the Problem

Use the medical references you have available to answer the following questions:

You have learned how to perform various laboratory tests in the course of this module.

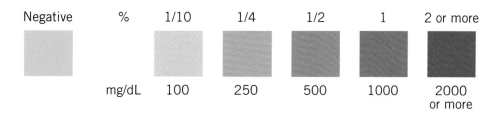

Negative   %   1/10   1/4   1/2   1   2 or more

mg/dL   100   250   500   1000   2000 or more

Figure 15
An example of a color chart that accompanies a packet of diabetes test strips

Kidneys rid the blood of excess water and waste products. Over the course of a single day, the human heart pumps about 1700 liters of blood through these organs—blood that carries essential molecules such as salts, glucose, and proteins as well as urea and creatinine, two waste products of protein metabolism. Eliminating waste products is relatively easy. Holding on to the essential molecules at the same time is the tricky part.

The kidneys accomplish their intricate balancing acts in thousands of tiny units called *nephrons* found in the cortex region. Under a microscope each nephron looks like a tangled array of tiny blood vessels and kidney tubules (see Figure 16). Filtration begins with a mechanical force in a part of the nephron called the *glomerulus,* a twisted mass of tiny capillaries. Of the 1700 liters of blood that flow through the kidney daily, about 180 are forced out through these glomeruli into the kidney capsules that enclose them. Included in this filtrate is essentially everything dissolved in the blood—good and bad. Cells and proteins, too large to pass through the porous walls of the capillaries, are left behind.

If this volume of fluid were immediately passed on to the bladder to be eliminated, the body would be depleted of water, essential nutrients and salts in three to four minutes. However, once the filtrate leaves the glomerulus, osmosis begins, steadily moving precious water molecules across the porous membranes and back into the blood. Other molecules, such as glucose and various salts can't get back to the blood by simple diffusion. These molecules are *actively transported* back into the surrounding capillaries. Special kidney tube cells move these essential molecules back toward the blood in a direction opposite to the one they would travel by simple diffusion. This critical flow of molecules takes both cellular energy and a set of unique proteins called enzymes.

As the filtrate completes its passage through the nephron, the blood vessels continue to pick up water and essential molecules. At the same time, the blood continues to discard more molecules to the kidney. Drugs and chemicals that were too large to pass through the glomerulus are actively excreted through the walls of the blood vessels.

The final filtrate, now called *urine*, contains water, some dissolved salts, toxins, urea, and creatinine. And instead of making 180 liters of urine in 24 hours, the kidneys pass only about 1.5 liters on to the bladder for elimination.

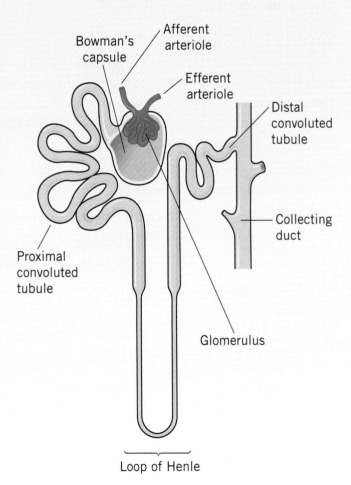

Figure 16
The kidneys contain thousands of nephrons, where the blood is filtered and the body's water balance is regulated.

? *Are there any circumstances in which a urine sugar level would be the most important test that is done?*

? *Are there any emergency medical situations in which knowing the glucose level is critical for making an immediate life-saving decision about treatment?*

Record your findings and discuss them with other members of your team. *note*📖

Laboratory 4 Bacteriology

Laboratory Overview

The only test in bacteriology that can produce reliable and useful results quickly is the Gram stain (Figure 17). Bacteria are classified by their shapes, groupings, growing requirements, and stain reaction.

- A spherical bacterium is called coccus (plural, cocci).
- A rod-shaped bacterium is bacillus (plural, bacilli).
- A curved or spiral-shaped bacterium is called a spirillum (plural, spirilla).

Each kind can occur in clusters, chains, pairs, or as singles. After Gram staining, if the bacteria appear red, they are Gram negative. If they appear purple, they are Gram positive.

Figure 17
Classifications of bacteria

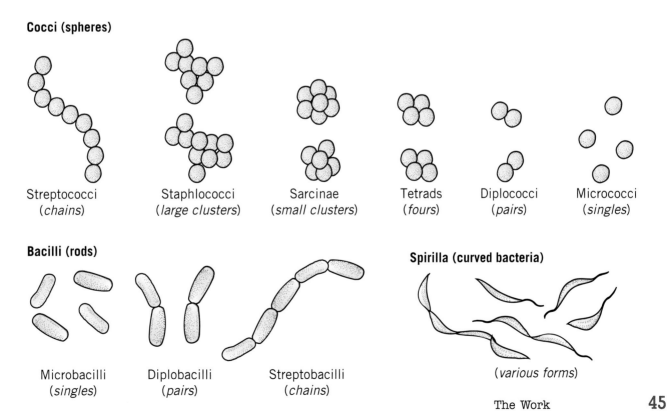

Cocci (spheres)

| Streptococci (*chains*) | Staphlococci (*large clusters*) | Sarcinae (*small clusters*) | Tetrads (*fours*) | Diplococci (*pairs*) | Micrococci (*singles*) |

Bacilli (rods)

Microbacilli (*singles*)   Diplobacilli (*pairs*)   Streptobacilli (*chains*)

Spirilla (curved bacteria)

(*various forms*)

When a patient is diagnosed with a bacterial infection, the first step before prescribing any treatment is to identify the specific type of bacteria involved. If standard swab and culture tests fail to make a positive identification, or if there is not enough time for these tests to develop, the technician must start with the most basic identification—the Gram stain test. This test classifies bacteria based on differences in the amount of a molecule called *peptidoglycan* found in the cell walls (Figure 18). Peptidoglycan, complex molecules that are part protein and part carbohydrate, function by strengthening the cell wall. Cells with peptidoglycan retain a purple dye during Gram staining and are called Gram *positive:* those with no peptidoglycan do not retain the dye and are called Gram *negative* (Figure 19). (In an actual laboratory, the stain will exhibit a range of colors.)

The amount of peptidoglycan in a cell not only determines the way that the cell retains certain stains but, more critically, also determines how the cells will react to antibiotics. Certain antibiotics—penicillin, for instance—are better suited for treating Gram positive bacteria because they interfere with the cell's ability to make peptidoglycan molecules. Without the strengthening molecules, the cell wall is weakened. Ultimately it breaks open, and the cell dies. Certain types of pneumonia, for instance, are treated effectively with penicillin. On the other hand, Gram negative bacteria are unaffected by an antibiotic like penicillin. Tuberculosis is one such disease. All this explains why the Gram stain test can be so important when deciding on proper treatment. Knowing the type of bacteria causing the infection is the first step in choosing the right antibiotic for treatment.

Bacteria are highly adaptable, and some carry genetic information that makes them resistant to particular antibiotics. One bacterium can pass copies of that information to others by means of a small circular fragment of DNA called a plasmid. If a patient is told to take antibiotics for ten days but stops after four days because he or she feels better, there is a good chance that enough bacteria carrying a plasmid with the resistant DNA will survive. The illness will return, but now most of the bacteria have adapted—they have acquired the resistant DNA—so the original antibiotic will no longer work. The patient will have to see the doctor again, and another drug will have to be used, often one that is stronger, but one that has more potential for causing negative side effects.

In recent years, the survival of resistant bacteria has caused the reappearance of many diseases that were once thought to be wiped out of our population. In large cities there has been an increase of drug resistant tuberculosis (TB). Some people cannot afford the cost of a 10-day supply of the antibiotic used to treat TB. Instead they often take only the samples the doctor

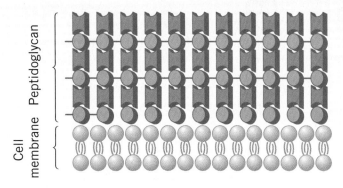

Figure 18
Peptidoglycan is a chemical component of the cell wall of some bacteria; Gram staining detects the presence of peptidoglycan.

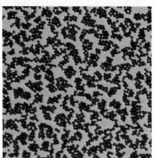 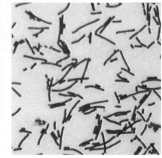

Figure 19
Gram-positive and Gram-negative bacteria

gives them, or they buy just enough medicine to make them feel better. Sometimes they lend or sell the leftovers to other infected patients, who also take a "feel better" number of pills. The partially treated individuals almost certainly become reinfected, but this time, the bacteria resist the standard treatment. What is more, others will also become infected with the resistant strain of TB bacteria. As the pattern repeats, the resistance of each succeeding generation of bacteria will increase. TB is now considered a multi-resistant disease—it is resistant to almost all of the drugs that once were used to treat it.

TB is not the only disease to develop resistance. Gonorrhea, a sexually transmitted disease that once was easily treated, has undergone a similar transformation. Research continues in order to stay a step ahead of these resistant varieties of bacteria by searching for and developing new antibiotics.

In this laboratory, each member of your team will perform a Gram stain and prepare the specimens for the day shift to set up for identification and sensitivity tests. Preparation of most specimens is done by ensuring that the specimen is properly packaged in a sterile container and labeled correctly, then placing it in the refrigerator. Some specimens may require immediate transfer to appropriate agar plates and incubation.

Materials

- protective gloves, goggles, and apron
- laboratory burner
- distilled water
- tap water in a wash bottle
- microscope with oil immersion objective
- a capped tube containing an unknown sample of non-pathogenic bacteria growing on a slant of agar medium
- nichrome transfer loop
- paper towels
- 4 microscope slides (one for every student in group)
- single dropper bottles of the following: crystal violet stain, Gram's iodine, decolorizer (95% ethanol), safranin stain
- immersion oil
- wire test tube rack
- shallow pan
- slide holder (hinged clothespin may be used)
- Clock with second hand or stop-watch

Safety

- Wear safety goggles, plastic gloves, and a laboratory apron throughout the procedure.
- Be careful when using the burner. Do so only as directed by your teacher.
- Tie back long hair and secure loose sleeves when working around a flame.
- Always check to ensure that the gas is off when you finish using the burner.
- Be careful when handling glass slides.
- Dispose of all materials as your teacher directs.
- Wash hands thoroughly with soap on completion of this laboratory experiment.

Procedure (for 3 or 4 team members)

1. Light the gas burner with a striker as directed by the teacher. Adjust the air vent on the burner until all traces of yellow or orange disappear from the flame.

2. Hold the nichrome wire by the attached holder. Place the wire loop in the flame until it glows red. (This is called *flaming* the loop.) Allow it to cool slightly.

3. Transfer a drop of distilled water to the center of a clean slide.

4. Reflame the loop, and let it cool.

5. Carefully remove the cap from the tube of bacteria. Flame the opening of the tube.

6. Using the wire loop, take a small sample of the bacteria and mix the sample in the water drop. Smear the mixture on the slide over an area the size of a dime.

7. Again flame the loop and the opening of the tube. Replace the cap and put the tube back in its original rack.

8. Hold the slide with the slide holder. Quickly pass the slide three to five times through the burner flame to evaporate the liquid. This technique, called *heat-fixing*, makes the bacteria stick to the glass *Do not* overheat the slide or it will break.

9. Place the heated slide across the top of a wire test-tube rack, smear side up, and allow it to cool. Place the rack in a shallow pan.

Figure 20

(a) Be generous as you flood the slide with crystal violet.

(b) Rinse away the excess stain with water.

(c) Blot the slide to dry it.

10. Flood the slide with crystal violet (Figure 20a), and let the slide stand for 1 minute.

11. To remove excess crystal violet, gently rinse the slide with tap water until the stream runs clear (Figure 20b). Tap the slide on a paper towel to remove any excess water. Replace the slide on the rack.

12. Flood the slide with Gram's iodine and let stand for 1 minute.

13. Use the wash bottle to rinse the slide. Tap the slide on the towels and replace it on the rack.

14. Rinse the slide with decolorizer, only until no more purple leaves the slide—about 30 seconds.

15. Use the wash bottle to rinse the slide. Tap the slide on the towels and replace it on the rack.

16. Flood the slide with safranin for 30 seconds.

17. Use the wash bottle to rinse the slide, gently blot it dry (Figure 20c), and replace it on the rack.

18. Repeat the procedure until each member of your group has made a slide of the team's assigned bacteria.

19. Observe the slide using an oil immersion lens. (See page 37 to review directions for using oil immersion.)

Analyzing the Data

Make a sketch in your laboratory notebook that shows the basic cell shape, groupings, and Gram stain reaction of the bacteria on your slide. Which of the following descriptions fits the pattern that you sketched?

- Gram positive cocci in clusters
- Gram positive cocci in chains

- Gram positive cocci in pairs
- Gram negative cocci in clusters
- Gram negative cocci in chains
- Gram negative cocci in pairs
- Gram positive bacilli in clusters
- Gram positive bacilli in chains
- Gram negative bacilli in clusters
- Gram negative bacilli in chains
- Gram positive spirilla
- Gram negative spirilla

Exchange slides with other teams in order to become familiar with a variety of bacterial types. Work together until every member of your team can make accurate identifications of the bacteria types available as samples in this lab. note📖

Arriving at Conclusions

Use medical references and Internet resources or other resource to determine which of these groups of bacteria are best treated with the penicillin family of antibiotics. What is the best alternative antibiotic therapy for someone who is allergic to penicillin? For which bacteria types is penicillin not indicated? ❓ note📖

If a smear reveals no bacterial presence in the specimen, would you expect the physician to prescribe antibiotics "as a precaution"? Read the Science Connection on page 46 to learn about problems that develop when antibiotics are inappropriately prescribed. ❓ note📖

Connecting to the Problem

You have now performed a variety of tests in your training to become a clinical laboratory technician. On the basis of your training and the medical resources you have available, if you were presented with several orders at the same time, in what order would you conduct the following tests? ❓

- stat type and cross match from the emergency room
- stat Gram stain
- routine blood glucose
- stat blood count with differential

Write your answers in your laboratory notebook, and justify the priorities you assigned. note📖

Laboratory 5 The Night Shift

Laboratory Overview

Having been trained to perform and prioritize a variety of laboratory procedures most commonly ordered on the night shift, your team is now ready work an overnight shift in the laboratory. Your task is to perform all of the tests ordered during your shift and to do so accurately and in the proper order. You must interpret the orders correctly and report the results appropriately. You must prioritize all samples according to the priority code and the department.

Materials

- protective gloves, goggles, and an apron
- request forms and simulated patient specimens
- CD-ROM
- your laboratory notebook, with records from all your training laboratories
- equipment to perform all tests in each laboratory area

Safety

- Wear protective goggles, disposable gloves, and a laboratory apron throughout the procedure.
- Be careful when using the burner. Do so only as directed by your teacher.
- Tie back long hair and secure loose sleeves when working around a flame.
- Handle hot containers with a clamp.
- Do not touch or spill any of the liquids.
- Dispose of used containers and materials as directed by your teacher.
- Wash hands thoroughly with soap on completion of all laboratory experiments.

Procedure (for 3 or 4 team members)

1. Gather your request forms and specimens from the central processing area.

2. As a team, prioritize the work, log in your notebooks how long it will take to complete, and decide who will cover which area of the laboratory. Your team may also decide to rotate tasks.

3. Run calibrations and controls on equipment as needed.

4. Run all procedures in the order you decided, and log all results in the log book.

5. Report any panic values directly to your supervisor so that the doctor can be contacted. Log any panic reports made.

6. Report all results on the request form, and route the form to the correct location.

7. When all of your work is done, double check your logs to be sure you left nothing out.

8. When you have finished checking your logs, organize the work and the reports you have made into a daily report to be presented to your supervisor and your class.

As you present your report be sure to

- justify the priorities assigned to your work

- explain which, if any, results were called as panic values

- provide data from your training to support your organization of the work

Presenting Your Findings

Your teacher will arrange a place for you to present your report. The length of your report will depend on the number of teams presenting and the amount of class time available.

Make copies of your results and supporting materials before the presentation begins so you can hand the materials to your teacher for reference during the presentation. Decide how your team members will share the roles in your presentation.

In your presentation try to use as many appropriate visuals as possible, including material from the CD-ROM and video. Make your presentation accurate, concise, well-supported, interesting, and visually stimulating.

Evaluation Criteria

Your presentation and report will be judged on the following criteria:

- The accuracy of your test results

- How well you followed the priority codes

- How well you organized your work

- The quality and clarity of your test reports

- The clarity, accuracy, and completeness of your presentation

Reviewing the Process

Reflect on the work you have done throughout your training program.

- How well did the training program prepare you to meet the requirements of working on the overnight shift in the clinical laboratory?

- What would you have changed about the training program to better prepare you to work in the clinical laboratory?

- How did you like working as a CLT? Is this something you might consider as a future career?

glossary

Agar: A powder derived from a marine alga used to gel a nutrient medium in a culture dish.

Agglutination: A clumping of cells that occurs when their surfaces are coated with antibodies.

Antibiotic: A drug used to treat bacterial infections, not effective against viruses.

Antibody: Protein produced by the body's immune system that attaches to surface features of foreign particles.

Anticoagulant: A chemical added to stored blood that prevents clotting. EDTA is an example.

Antigen: Surface feature of a cell or particle that stimulates the immune system to produce antibodies. Important in identifying cells as self or foreign.

Anti sera: A cell-free derivative of whole blood containing antibodies specific for a certain antigen.

ASAP: Priority code that means "as soon as possible."

Aspirate: To draw up by creating a partial vacuum; to suck.

Bacillus: Rod-shaped bacterium.

Bacitracin: A antibiotic obtained from a strain of Bacillus bacteria. Effective in controlling some penicillin-resistant strains of Gram-positive bacteria.

Bacterium (sng) **bacteria** (pl): Single-celled microorganisms with cell walls and lacking nuclei; classified by shape, groupings, staining properties, and food source.

Blood type: Classification of red blood cells based on the presence or absence of certain surface antigens.

Calibrate: Adjust the performance of an instrument so that the reading for a standard material of known properties is correct.

Centrifuge: Laboratory instrument for separating the components of liquid mixtures based on density properties. Spinning generates layers in the test tube with heavier elements at the bottom and lighter ones at the top.

Coccus: A bacterial cell shape that is spherical.

Control: Sample with known identity and properties. Tested to ensure that a clinical test procedure is giving valid results.

Cross match: A test performed prior to a transfusion in which a sample of donor blood is mixed with a sample of recipient blood to check for an agglutination reaction that would signal incompatibility.

Culture: A laboratory population of a type of microorganism that grows under ideal conditions in a closed container of nutrients selected for encouraging rapid growth.

Diabetes: A disease characterized by high levels of blood glucose caused by the failure of the pancreas to produce adequate amounts of the hormone insulin.

Differential Cell Count: A tally of the percentages of certain cell types observed on a slide. Generally applied to the different categories of white blood cells.

EDTA (ethyleneadiamine tetraacetate): An **anticoagulant.**

Electrolyte: An ion present in a solution that allows an electrical current. In blood, the most commonly measured are sodium (Na^+), potassium (K^+), and chloride (Cl^-).

Endocrinology: The science and technology focused on the body's chemical messengers or hormones and the various tissues and organs that secrete them.

Gram positive, Gram negative: A staining property of bacteria cell walls. Gram positive retain a purple dye when washed with a solvent; Gram negative do not.

Gram stain: A staining technique in which a fixed smear of bacteria is treated with a series of stains and washes. The final color is important in classifying the bacteria.

Hemacytometer: A microscope slide etched with a grid used for counting blood cells.

Hematology: The technology related to the analysis of blood.

Hemoglobin: The molecule in red blood cells that carries oxygen from the lungs and releases it to cells.

Hemolytic transfusion reaction: Life threatening condition resulting from transfusing mismatched blood into a patient. Antibodies in circulation for the foreign antigen set off a chain of reactions in which the immune system breaks apart (destroys) the foreign red blood cells.

Heparin: An anticoagulant.

Incubate: To store a culture under specified temperature and humidity conditions for a period of time.

Insulin: A hormone secreted by the pancreas that results in lowering the glucose concentration in the blood.

Ion selective electrode: An instrument used to measure the concentration of certain ions in a solution.

Kidney: Organ that filters the blood and excretes waste.

Microbiology: The study of bacteria, viruses, and other organisms too small to be seen without a microscope.

Nichrome: A nickel-chromium alloy with a low heat capacity and high melting point. Wires made of this alloy are useful in laboratory transfers of living cells because they cool rapidly after flame sterilizing.

Osmotic balance: A condition in which the amount of water leaving a living cell through the cell membrane is equal to the amount of water entering.

Pancreas: An organ near the stomach that secretes digestive enzymes and insulin.

Peptidoglycan: Chemical component of cell wall of bacteria. Its presence or absence in bacteria cell walls determines the Gram staining characteristics. Also determines its sensitivity to a particular antibiotic.

Petri dish: A flat dish with a loose fitting lid designed to culture organisms on an agar-based medium under sterile conditions.

Phlebotomist: A technician who specializes in drawing, labeling, storing, and delivering blood samples.

Plasma: The watery fluid of the blood, which contains a variety of proteins and fats, dissolved salts (electrolytes), and glucose (blood sugar) and in which blood cells and other structures are suspended.

Plasmid: A sequence of DNA that is taken in by bacteria and incorporated into the genetic code. May carry genes for drug resistance.

Proteins: Nitrogen-containing organic molecules made of long chains containing as many as 20 different amino acids. Specific composition determined by DNA codes. Found in all living things.

Reagent stick: A prepackaged strip to which small samples of test chemicals are attached. Designed to be dipped into a liquid sample. Color changes indicate the presence or absence of substances.

Red blood cells (RBCs): Also called *erythrocytes;* red-colored blood cells that have no cell nuclei; contain hemoglobin, which is responsible for transporting oxygen molecules throughout the body.

Rh group: Another system of antigens on the surfaces of red blood cells. Important in blood typing.

Rhesus factor: Also called Rh factor. An antigen found on human red blood cells that is either present (Rh+) or absent (Rh−). Important in blood typing. First discovered in rhesus monkeys.

Routine procedures: Procedures that are conducted to assess general health but are not considered specific to certain diagnoses or certain problems.

Saline solution: Aqueous salt solution in which the ion concentration matches that of body fluids.

Spirilla: Curved or spiral-shaped bacteria.

Standard: An aqueous solution of known concentration used for calibrating an instrument.

Stat: A priority code meaning that the results of the test are critical for patient care, and are needed immediately.

Streptococcus bacteria: Spherical bacteria that occur in chains. Feature used in diagnosing infections.

Toxicology: The science and technology related to the detection and study of substances that interfere with normal body functions.

Transfusion: Procedure in which blood from one individual is transferred to another.

Urinalysis: Testing of a urine sample to determine its chemical composition. Used to diagnose disease and to detect the use of certain drugs.

White blood cells (WBCs): Blood cells that are typically larger than red blood cells. Five basic kinds of WBCs predominate: *basophils,* which release chemicals like histamines to control blood vessel dilation; *neutrophils* and *monocytes,* large cells that engulf and eat foreign material; and *eosinophils,* which engulf certain parasites and control the body's allergic reactions.

additional resources

American Society for Clinical Laboratory Science (ASCLS)
7910 Woodmont Avenue
Suite 530
Bethesda, MD 20814
phone: 301-657-2768
fax: 301-657-2909
http://www.ascls.org/ preeminent organization for clinical laboratory science practitioners.

Center for Disease Control and Prevention (CDC)
1800 Clifton Road, Atlanta, GA 30333
phone: 404-639-3311
www.cdc.gov lead federal agency of HHS for protecting the health and safety of people in the United States with news, data and statistics, health topics A–Z.

Clinical Laboratory Management Association (CLMA)
989 Old Eagle School Road, Suite 815
Wayne, PA 19087
phone: 610-995-9580
fax: 610-995-9568
www.clma.org international organization with members who are responsible for laboratories and clinical services in hospitals and health-care networks, group practices, and independent settings.

National Library of Medicine (NLM)
8600 Rockville Pike, Bethesda, MD 20894

MEDLINE*plus*
www.medlineplus.gov premier consumer website on health information from the National Institutes of Health.

Rush University
College of Health Sciences
Chicago
http://www.rushu.rush.edu/medtech/ information on undergraduate and postgraduate training to become a clinical technician.

U.S. Department of Health and Human Services (HHS)
200 Independence Avenue SW, Washington, DC 20201
phone: 877-695-6775
www.hhs.gov principal government agency for protecting the health of all Americans with governmental policy and affairs, news, and links to related web sites.

credits

Cover photos:

chip: David C./CORBIS;
pills: Jean-Pierre Lescourret/CORBIS;
vials: Tek Image/Science Photo Library/Photo Researchers, Inc.;
police tape: PhotoDisc, Inc.;
cows: PhotoDisc, Inc.;
pipeline: SuperStock, Inc.

p. i (bottom and repeated throughout):   PhotoDisc, Inc.

p. ix (top and repeated throughout):   Science Pictures Limited/CORBIS.

p. 15 (top):   Microworks/Phototake.

p. 15 (bottom left):   Institut Pasteur/Phototake.

p. 15 (bottom middle):   Dennis Kunkel/Phototake.

p. 15 (bottom right):   Eye of Science/PhotoResearchers, Inc.

p. 16:   Dennis Kunkel/Phototake.

p. 17:   Pictor International, Ltd./PictureQuest.

p. 23 (left):   Yoav Levy/Phototake.

p. 23 (right):   Lynn Goldsmith/CORBIS.

p. 42 (right):   Rick Lance/Phototake/PictureQuest.

All other photos: American Chemical Society/Take One Video.